Faux Real

Genuine Leather and 200 Years of Inspired Fakes

Also by Robert Kanigel

High Season

Vintage Reading

The One Best Way

The Man Who Knew Infinity

Apprentice to Genius

Faux Real

Genuine Leather and 200 Years of Inspired Fakes

Robert Kanigel

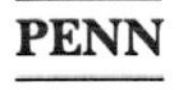

UNIVERSITY OF PENNSYLVANIA PRESS

PHILADELPHIA · OXFORD

First published 2007 by Joseph Henry Press

Paperback edition published 2010 by
University of Pennsylvania Press
Philadelphia, Pennsylvania 19104-4112
www.upenn.edu/pennpress

Printed in the United States of America on acid-free paper

10 9 8 7 6 5 4 3 2 1

ISBN 978-0-8122-2132-9

One little room, an everywhere

Contents

Part I

Imitating the Inimitable

1
Material World

It was 1963 and they were offering a new material to the world.

No single enterprising hero was behind it, no one glib-tongued salesman; it was *they*: Men like Johnny Piccard, a young chemist, scion of one of America's early ballooning families, his deep voice and fertile ideas booming through the halls of the Experimental Station. And Charlie Lynch, the boyish-looking marketing director, plucked from a job selling vinyl in Detroit. And Bill Lawson, who'd flown B-17s over Germany during the war and now was in charge of getting the new material into American shops and homes.

Theirs was a material no caveman had ever touched, no European monarch had ever worn. It could never be found beside a rocky Appalachian trail or high up in the jungle canopy of the Amazon or buried deep in a Siberian mine. It was not iron, bronze, or stone. Not silk, linen, leather, or wood. Those were old and familiar, the stuff of common speech: *Smooth as silk. Nerves of steel. Hell bent for leather.*

This new material, years in the making, the creative and technical achievement of hundreds, boasted properties that had never before existed together and promised to reach into the lives of millions. To listen to Du Pont—the "they" who'd created it—it would make those lives better.

It was October 1963, and Bill Lawson stood before dozens of reporters from trade, consumer, and fashion publications at the New York Hilton Hotel in New York City. He was there, he said, to "celebrate the coming of age of a wholly new, chemically created material." Lean, trim, 6-foot-2, an automobile aficionado with a barn full of cars out in the country, Lawson was a 45-year-old chemical engineer out of Cornell who'd joined Du Pont back in 1949 and risen through the ranks. He would try "to maintain a seemly restraint," he said, but hoped he might be forgiven for any "excesses of enthusiasm" slipping through. "We announce today," he said, "a long, long dream come true. And the pursuit of the dream has been for us, and the team after team of men who followed the dream before us, a kind of chemists' *Pilgrim's Progress.*" Lawson was comparing the development of the new material to John Bunyan's classic 17th-century allegory. He was likening Du Pont's efforts to a virtuous man's pilgrimage through life.

The year 1963 was a time nearer in spirit to hula hoops than love beads. It was only later, in retrospect, that the assassination of President Kennedy in November, the arrival of the Beatles in America, the publication of Betty Friedan's *The Feminine Mystique,* and America's descent into the southeast Asia war would come to seem a watershed in recent American history; the *real* "Sixties," with their questioning and protest, came only later. The early 1960s were more like the 1950s; they packed more of the technological exuberance of General Motors' Futurama exhibition at the 1939 World's Fair than of the cynicism nurtured by Vietnam body counts. In 1963, in St. Louis, the Gateway Arch was rising over the Mississippi, celebrating the opening of the American West to exploration. In New York City, the twin towers of the World Trade Center were going up.

Bill Lawson, then, spoke from the *other* side of a great historical divide, when miracle products could still be trumpeted to an eager and accepting public. It was a time when engineers formed a kind of techno-avant-garde, and when the prospect of human beings on the Moon, as President Kennedy had announced was the nation's goal in 1961, evoked wonder, had not yet ossified into grainy, half-remembered black-and-white video images radioed back from space. Now, set against this sunny technological backdrop, Lawson and company were bestowing on the world a new material that could become as ubiquitous as the clay in dishes, the paper in books, or the nylon in stockings. His talk, and the whole Du Pont campaign, gave off an icy white glow of inevitability.

At the New York Hilton that day in 1963, Bill Lawson described the culmination of research that went back more than 25 years. Du Pont's new product was called Corfam. It was supposed to replace leather.

Leather was as old as humanity, early evidence of its use going back to Egyptian wall paintings and the Bible. It was an icon of naturalness and authenticity, a symbol of luxury, its smell and touch stirring primeval feelings of sensual pleasure. "A product of Nature of wondrous structure and beauty," one Englishman called it. Scientists gushed over the baroquely intricate fibrous patterns that lay beneath its grained surface. Diderot and D'Alembert devoted a volume of their *Encyclopédie* just to the special tools that, *circa* 1760, went into its making and using. Whole glossaries were needed just to keep track of its numberless variations: Cordovan, made from horsehide; chamois, soft oil-tanned leather made from sheepskin; shagreen, an abrasive sharkskin; morocco, red goatskin leather from north Africa. Dye it black, cut and shape it into a jacket draped with zippers, buckles, and belts, throw it around the shoulders of Marlon Brando in *The Wild Ones,* and you had an enduring statement of youthful rebellion. Fashion it into calf-hugging stiletto-heeled boots and you had the makings of a full-scale fetish object. Pull it snugly around a wooden last, tack it down, and shoemakers would rhapsodize about its virtues. Leather stirred superlatives. It aged gracefully, grew more beautiful with time. "There's nothing like leather" was the mantra every self-respecting tanner asserted.

But then, one Friday evening in October 1961, Fred O'Flaherty came to Salem, Massachusetts, to say that maybe there *was* something like it, something that threatened this age-old material and the men who made it.

O'Flaherty was one of the grand old men of leather. A World War I veteran, holder of veterinary and medical degrees, and director of the research lab of the Tanners Council, he'd spent his life helping tanners make better leather, garnering awards along the way from just about everyone in the business with one to give. He was a man for whom the skin of a cow was as familiar as his own, an otherwise no-nonsense midwesterner not given to fits of eloquence who once said he'd never met a soul who didn't like leather. And now, at the age of 68, he'd come to the old Hawthorne Hotel in Salem, Massachusetts, a town that, long past its witch-trial notoriety, had become, with neighboring Peabody, the center of the New England leather and shoe industries. It was the New England Tanners Club's opening meeting of the year, and O'Flaherty was the guest speaker. The treasurer's report was given, new officers inducted, O'Flaherty introduced. The title of his talk was "The Invasion of the Tanning Industry."

To invade, said the dictionary, meant "to infringe or encroach, to enter for conquest or plunder," and that, said O'Flaherty, was precisely what tanners faced, in this case by that "common enemy of every tanner, substitutes." Materials that looked like leather, or served for leather, but weren't leather. Already 7 in 10 shoes made in the United States had soles made from nonleather compounds, like Neolite, a hard rubberlike material; *that* battle was long lost. And now shoe uppers—the stylish part of the shoe you buffed until it gleamed, that conveyed first impressions to the world, that showed off a dancer's slim ankle or presented a man as serious, prosperous, and important—these, too, might now be made from nonleather materials.

O'Flaherty had been warning of synthetic threats for two decades. Before World War II, when Goodyear brought out Neolite, he'd gotten his hands on some and alerted the Tanners Council. But just then, as he recounted later, the tanners were selling every scrap of leather they made and couldn't be bothered to listen. By war's end, Neolite was well established. By 1960 it was being made into 400 million pairs of soles per year, and old-line tanners specializing in sole leather—American Oak, Zellers, Laub—had been driven out of business. Now, O'Flaherty had been saying for at least a year, tanners faced a new menace, from a material that, made into shoe uppers—by far the largest part of the business—even old leather hands couldn't always distinguish from the real thing.

"It is common talk today," he reminded his Salem audience, "that a material made by the Du Pont Company"—it didn't yet have a name—posed "a real threat to the shoe upper tanner." Du Pont's and other synthetic materials, he warned, were "in a stage of development which corresponds to the status of Neolite less than 10 years ago." During the discussion period, O'Flaherty hauled out his traveling display of leather substitutes. One was a shoe made from the Du Pont material.

From the late 1950s all the way up to the introduction of Corfam in late 1963, it was like that. Trade shows riddled with rumors. Panel discussions at conventions. Tanners beating their breasts. Cassandra-like warnings alternating with calls for calm. Anticipation. Dread. A kind of fever, of a great moment impending.

In 1960 there came a seeming reprieve. From Du Pont headquarters in Wilmington, Delaware, it was reported that the "revolutionary new shoe upper material may still be two and a half years away"; its name was not yet known, but "Cavalon" and "Dolyn" were among those mentioned.

Then, a rumor: A shoe maker was supplying the Navy with men's oxfords "made entirely without leather. The shoes, intended for wear testing, employ Du Pont's new upper material."

November 1961, Milwaukee: The Du Pont material, the Tanners Council of America heard, was said to be "far superior to any of the synthetic 'substitutes' of the past and present."

And so, in April 1963, with Du Pont's new synthetic being tracked like a hurricane rampaging up the Eastern Seaboard, it was a matter of no slight interest, worth a headline of its own in *Boot and Shoe Recorder*, when the announcement finally came from Wilmington:

DU PONT'S NAME FOR UPPER: CORFAM

If you were in the industry back then, you couldn't get away from Corfam, the threat it represented, the revolution it implied. *Leather and Shoes* chronicled its coming in telegraphic-style reports. "Du Pont Material Scores Again. . . . Build-up of new Du Pont upper material gains momentum." *Boot and Shoe Recorder* convened a Tanners Panel of 60 top-level tanning executives, quizzing them about "their greatest challenge"—man-made upper materials. A division of W. R. Grace, weighing its own synthetic shoe upper, collected press clippings and patents devoted to nonleather substitutes, compiling a report, "The Tanning Industry and Artificial Leather," that focused mainly on Corfam.

In 1962 a Du Pont executive, William E. Kreuer, showed up at the Delaware Valley Tanners Club, his appearance doubling the normal turnout to 200. It was the first time Du Pont had dared appear before tanners and their suppliers; Kreuer allowed that inviting him was "like asking Krushchev," the Soviet Communist leader, "to talk before the Knights of Columbus." At the end, he and marketing manager Charles Lynch showed off men's, women's, and children's shoes made from the new material. Tanners gathered around, by one account, "with all the anxiety of tots on Christmas morning," fondling, twisting, squeezing, and smelling.

"Not bad," said one.

"Feels like plastic," said another.

"Nope, it feels like leather," said a third tanner, his voice, by one account, seeming "a trifle sad."

At their 1963 annual meeting, soon after Corfam's introduction, the tanners heard an advertising industry executive lambaste them:

> Your record of sacred cow worship, of ultraconservative planning, does not offer reason for optimism. . . . Your industry has been a laggard in technological advances. You have invested a smaller percentage of sales in scientific exploration than many other industries. You have shown less of an inclination to explore challenging new scientific vistas . . . and you permitted your new competitors to make the major advances in man-made leathers. Some members of your industry are still trying to wish man-made leathers away.

It was as if the tanners had invited a dominatrix to mercilessly and publicly humiliate them.

It was a demoralizing time for the tanning industry, those early 1960s. As one trade journal put it, "the public relations department of America's largest chemical manufacturer" had succeeded so brilliantly in ushering in the synthetic future, that old leather hands had come to see in Corfam "qualities even the most ebullient on the staff of its producers had yet to think of." The tanners stewed in self-pity and lamentation. "It's time," said the article, that "the industry stopped trying to act a pallbearer at its own funeral."

Occasionally someone did try to dispel the gloom. Why, three-quarters of the people in the world had never owned a pair of leather shoes, it was said. Global living standards were rising; leather would benefit. The wool industry had survived synthetic fibers and now found itself healthy and robust. Likewise for cotton and silk. So would leather. One trade journal editor weighed in with a dollop of happy news: "Just when the idea of synthetic leathers had the tanners trembling in their hides, fashion did a complete turnabout and decreed that the Natural Look Is In . . . and the ONE look in leathers that can't be synthesized yet is the natural leathery look."

It was leather itself, and its unique qualities, to which leather's defenders clung. "Our modern world has seen the entry of many new 'miracle fibers,'" observed another trade journal around the time of Corfam's introduction. "But significantly, the most miraculous of the miracle fibers is still the oldest of fibers—leather. Its real miracle lies in the fact that despite the ceaseless probings of the scientists and chemists, mystery still shrouds [its] complex and fascinating structure."

A few months after Du Pont shook up the Delaware-area tanners, one of their number, Arthur J. Edelman, a tannery vice-president, came before them to deliver a pick-me-up, asserting, "I'm in love with leather." The tanners had to be flexible. "Beat the synthetics at their own game," he said. "If kids demand sneakers, fine, make leather sneakers."

"We cannot afford to surrender," the president of the Tanners Council told 500 tanners in Milwaukee. "We cannot afford to imitate imitations."

All the rah-rah could seem reminiscent of a last-place team trying to rev itself up to play the Yankees. Still, they tried. In 1963, a New Jersey tanner presented retail seminars touting leather's advantages to shoe salesmen gathered in Chicago. He outlined the principles of tanning, pointed up leather's virtues, even brought in a podiatrist to warn of the dangers of synthetics. "All the plastics and all the materials that are being tried to simulate nature," he said, just didn't measure up. Why, newly identified skin ailments, he warned ominously, could be traced to nonleather footwear.

"There's nothing like leather," said the ad-budget-tripling Leather Industries of America, warning consumers of the danger of putting anything on their feet that wasn't. "Naturally, you prefer leather," ran the headline to an ad appearing in *Life* magazine in September 1963. "That's why they are trying to imitate it."

> Consider the things you can count on in real leather shoes: their natural breathing and flexing action; their cooling absorption of normal foot moisture; their ability to hold their shape. These are benefits imitation products try to duplicate.
>
> Why accept imitations when leather shoes give your family the protection and deep down comfort they need—naturally.

But leather was of the past. When Corfam finally appeared in a paroxysm of product introductions in 1963, and then early the following year in four-page, full-color spreads in *Vogue* and *Elle*, on television, in big-city department store ads, and at the New York World's Fair, Du Pont's message was simple: It invited the public to

STEP INTO TOMORROW

Place, into a small container, some dry absorbent material, like silica gel, whose weight you know down to the last milligram. At the container's neck, attach a test membrane of Corfam, so that the gel is cut off from the outside—cut off, that is, save for whatever might pass through the test material. Let the apparatus sit there, in a chamber controlled for temperature and high relative humidity. Then, after a few hours, open it up and weigh the silica gel. It will be heavier, a little wetter. "Water vapor permeability" is what the chemists are measuring, and it's what you want in a shoe, especially on a hot summer's day.

Corfam "breathes." Vinyl doesn't; leather does.

Corfam breathed like leather.

Corfam looked like leather, bore the imprint of leather's distinctive grain. And it was to be used for products normally made of leather. But it had never had blood-engorged capillaries run through it or bristly hairs erupt from it. Like most plastics, it was a product of oil pumped up from the bowels of the earth and made into polyester and polyurethane; then melted, spun, matted, shrunk, pricked, and otherwise processed in a precise choreography of chemical and mechanical steps. Corfam was no tacky low-life vinyl. It was protected by a slew of patents covering every stage of its manufacture; there was one for making dense, fibrous mats that mimicked the dense, fibrous tangle of leather's flesh side; another for coagulating a polyurethane film in water so that it formed fine microscopic pores. Samples of the stuff had been made into shoes and tried out on hospital nurses, beat cops, and rough-housing street kids. Du Pont had pumped millions of dollars worth of engineering and marketing savvy into it. The company's preeminent chemical innovation, nylon, hadn't existed before Du Pont made it—and then, a miracle, it did, forever changing modern life. Likewise Corfam, already being called "the next nylon."

But, to state the obvious, this new miracle material that was intended to breathe like leather, do the work of leather, and be better than leather *wasn't* leather.

It was leather*like.* It was synthetic leather. Artificial leather. Imitation leather.

Fake leather.

Or, to use the French word that slyly slips around the stigma of fake, *faux* leather.

Du Pont would try never to mention Corfam in the same breath as leather. It pictured its product as unique, *sui generis*—neither leather, nor plastic, nor a chemical, nor anything but itself. Press releases issued at the time of its introduction included a note advising that Du Pont "avoids such terminology as 'man-made leather,' 'artificial leather,' or 'leather replacement' in describing Corfam."

But in fact, across all the years of its development, it had been leather, and leather virtually alone, that Du Pont had in mind. Back in 1937, according to one internal document, the company had surveyed the leather goods industry to weigh the potential of "synthetic polymeric materials to compete with leather." In 1949 the aim of the early research that would culminate in Corfam was, Du Pont chemist John Piccard recalls today, "to make it more like leather than any other existing thing." And in the mid-1950s, Du Pont's whole edifice of patent protection served innovations aimed, in the words of one of its patents, at achieving "the appearance of leather combined with the properties of leather." It was leather that was mentioned in early Du Pont research memorandums, leather whose physical and chemical properties Du Pont engineers and scientists sought to emulate, leather that provided almost the only yardstick for measuring Corfam's success, leather that Corfam was out to beat.

To its Experimental Station on the bluffs above Brandywine Creek outside Wilmington, Du Pont had for half a century brought sharp technical minds to develop new textiles, adhesives, paints, fibers, and films. These new products the company pushed and promoted with legendary tenacity, protecting them with unbreachable patent defense works. *Better Things, for Better Living . . . Through Chemistry.* That was the Du Pont slogan, and that was Corfam's promise, too. No-scuff, keeps its shape, and all the rest: *Better living.* "We see at present," said Bill Lawson to the journalists gathered at the New York Hilton in 1963, "no technical bar to man-making Corfam across a range of appearance from armor-like crocodile hide or the most fragile, soft chamois." There was nothing they'd be unable to do. Women would wear Corfam pumps to charity luncheons, men would wear Corfam golf shoes on the fairways, and yet another age-old material would fall victim to the implacable forces of technological progress.

It is not often that we see this happen, where something we use every day, and think of as being made from a particular material, changes forever, before our eyes; that, if Du Pont had its way, was what Corfam meant.

Shoes were as intimately bound up with one material, leather, as any pairing of object and material you could imagine. "An outer covering for the human foot, normally made of leather. . . ." That's a shoe, according to the *Oxford English Dictionary.* Shoes have been made from fabric, polyethylene, and wood, ornamented with rhinestone, enamel, and brocade; in *La Dolce Vita,* Anita Ekberg frolicked in the Trevi Fountain wearing satin stilletos. But until the coming of sneakers with their rubberized canvas uppers, those were the exceptions. Roman sandals? Indian moccasins? Leather. Likewise the haute couture slippers of Milanese designers or the brogues of fine English bootmakers. Even as late as 1963, 80 percent of all American leather went into shoes.

It was this age-old identity, this near equivalence of SHOE and LEATHER, that Du Pont attacked with Corfam. Later, other products would be made from Corfam, like purses and golf bags. But it was shoes that had inspired the research investment in the first place; of the 600 million pairs of shoes made in the United States in 1961, the leather ones among them each required one and a half square feet of hide. *Do the numbers.* Even back in 1937, when Du Pont researchers were starting to think about synthetic alternatives to leather, it was shoe uppers that had been identified as the biggest market by far, leather's virtues *as foot covering* those that any new synthetic material would need to match.

If Corfam successfully broke that cord of connection between object and material, it would represent the intrusion of a synthetic material into a personal human realm, the foot, that was peculiarly intimate and idiosyncratic. With its curled toes, its bony protuberances, its calluses and soft moist places, the foot was as individual as a fingerprint. The shoe that covered it was an icon of individuality, a space inhospitable to all but that one proper foot. To put yourself in someone else's shoes is to inhabit another's life—and so virtually impossible. Plutarch tells of a Roman assailed for divorcing his seemingly blameless wife; by way of answer, he displays his shoe which, though finely made, pinched and tortured him in ways his friends could only imagine. The smells and secretions of the shoe-shod foot can make it as personal as the underarms or genitals and give it, for some, almost as erotic a tang. And it was into this intimate human space, so long the realm of a single natural material, that Corfam presumed to intrude.

To the tanners and to Du Pont, the coming struggle was a matter of livelihood—a tannery closed in New England, maybe, or a chemical plant opened in Tennessee, more income to shareholders, or less, personal prosperity enjoyed in one generation, just another historical factoid to the next. But viewed through another lens, this marketplace battle was part of a larger struggle between traditional and modern materials. Five thousand years of sail made from natural fibers like cotton and hemp; 2,000 of windows and glass; a hundred of bicycle frames and steel. These and many other of the old familiar links between object and material have been sundered—by sails of Dacron, shop windows of Lexan, bike frames of titanium. And our lives have changed as a result—in small, barely perceptible ways, sometimes at the very edge of awareness, yet cumulatively large. Was just such a transformation taking place in 1963? Would we wear shoes of leather, or synthetic leather? Depending on the answer, everyday moments in the lives of millions stood to change. In the balance stood the feel of shoes on the foot, the experience of slipping them on, of cleaning and polishing them, of how they endured the maltreatment of daily life, of how they looked to the eye and felt to the hand. All those trifling experiences of eye, nose, and fingertip—small pleasures, petty annoyances—were being influenced right there, on that gritty commercial battleground.

In her 1984 song "Material Girl," Madonna reminded us that we are "living in a material world." The song meant "material" in the acquisitive and worldly sense—possessions, wealth, comfort, the good life. But in a broader sense, all people inhabit a material world—of the things and objects we use every day and that have their own distinctive textures, weights, finishes, and feels.

These materials have changed.

On September 19, 1991, a German couple hiking high in the Tyrolian Alps in northern Italy stumbled upon a body jutting out from a melting icepack. *A mountaineering accident victim?* Erika and Helmut Simon wondered. But soon archeologists and other scientists determined that the body was almost 5,000 years old, miraculously preserved in the ice since its death. The belongings of this Iceman, as he became known, were almost perfectly preserved as well. Around his waist he wore a belt pouch made of calf leather within which were found implements of flint, an awl made from bone, and a piece of tinder, probably tree fungus. He wore a fur cap, to which two leather straps were fitted. He wore an upper garment made of furs stitched together; leggings of light brown fur; a loin cloth of leather scraped thin; a long cloak plaited from grass; and shoes with cowhide soles, insulated with a thick layer of grass, and fur uppers, attached to the sole, forming a kind of boot.

Early men and women, like the Iceman, had raw materials such as rock, wood, and animal hide all around them; these they shaped, pounded, shaved, salted, dried, and sewed to better use them. From bugs came shellac. Reeds formed baskets. Clay went into fired pots. Metals like iron, bronze, and tin weren't so different; fire extracted them from their ores. Fermentation transformed a pot of hops into beer, a barrel of grapes into wine. One material fairly dissolving into altered form but never so remote from the original that you couldn't feel its origins in nature.

But day by day, these old materials, with their roots deep in human history, have given way to new ones. Materials variously better, stronger, glossier, easier to work, or cheaper than the natural ones. Acrylics, not oils, in our paint. Cabinets made of pigmented plastic, not wood. Books bound in vinyl, not leather or linen. Food made in factories. Fragrances cooked up in laboratories, not distilled from flower petals. One by one they're introduced, make a mark on the world, and are soon taken for granted. An earthen pot becomes a galvanized steel pail, the steel pail a plastic tub. Gradually, the old materials disappear and 21st-century human beings come to inhabit a new world, of new materials.

Some years ago the noted photographer Paul Menzel performed an audacious artistic and cultural experiment, one later embodied in a book, *Material World: A Global Family Portrait.* He and 15 other photographers went around the world—to Mali, Kuwait, Cuba, Bhutan, Bosnia, the United States, 30 countries in all—and in each prevailed on a family to collect all their worldly goods and set them out in front of their house to be photographed, in one formal tableau, along with the whole family. Some of the contrasts, of course, were entirely predictable: the desperately poor Haitian family, arrayed outside their shack with a thin assortment of pots, buckets, bare wood chairs, and crude farming implements; the well-off American family from Pearland, Texas, with three radios, three stereos, five telephones, toys, cars, appliances, multiple thises and thats, a proud glut of abundance ranging across the two-page spread.

But for all these differences, there was much in common, too. Modern materials had infiltrated the lives of every culture. Plastic tubs could be seen among the Haitians. The Guatemalan family, its hovel decorated with plastic flowers, was shown piled onto bus seats that looked as if they could be Naugahyde. The Indian women wore what appeared to be plastic spangles. An Albanian child wore plastic boots adorned with a Teenage Mutant Ninja Turtles design. A Mongolian family, who lived on the outskirts of Ulaanbaatar in a *ger*, the traditional tentlike house, claimed as their most valued possessions a wooden statue of the Buddha and their television.

True, the walls of the Ethiopian house were plastered in cattle dung and the Vietnamese family sat on a bamboo bench. But Menzel's project tells us that the old, familiar standbys—iron, wood, stone, leather, wool, straw, mud—are giving way, around the world, to more processed materials with a shorter history at humanity's side, especially plastics in their various guises. Each day the material world changes a little. Each day, in a slow yet seemingly inexorable process, natural materials yield to synthetic ones; handmade objects to machine-made ones; rough, irregular surfaces to plain or glossy ones; diverse and variegated things to nearly identical ones. Such changes normally occupy the background of our lives, not the foreground, and can seem as harmless and inevitable as the rising of the sun, unworthy of note. But in the end, after 10,000 years of human civilization, the primeval essence of the Serengeti Plain gives way to the bright plastic sheen of a New Jersey shopping mall.

Yet the new materials often give a nod to the materials they replace. Thin aluminum siding on suburban houses impressed with a wood grain. Coarse stucco slathered onto brick to resemble stone. The thin veneer on a midpriced bookcase intended to suggest in its few thousandths of an inch something of the richness of walnut. The vinyl seats in an airport waiting room bearing the telltale surface pattern of leather. In the evidence of such material sleight of hand, it can seem as if stone, leather, and wood exert some atavistic hold on us that new materials somehow feel bound to acknowledge, like a newly elected politician who makes flattering, if insincere, utterances about the virtues of his predecessor. But just as easily, we can see the new materials all tarted up for show as simply dishonest, not cheerfully *faux* but crudely false, thus contributing to the sense of deception and unreality that can seem the hallmark of our age.

Over the centuries, artisans, engineers, and inventors found ways to put exotic veneers on low-grade wood, to make it look beautiful and rare; to plate base metals with gold so that they looked as if they *were* gold; to make molded plaster pass for fine woodcarving. Later, with some of the early plastics, they imitated such coveted, expensive materials as ivory, ebony, and tortoiseshell and did it so well they looked for all the world like ivory, ebony, and tortoiseshell. They imitated diamond so well that only a jeweler could tell the difference. They made photocopy machines that could counterfeit dollars and pounds. They made plastic fibers that looked like silk.

Faced with a traditional process, product, or material, one whose roots often went back to the earliest civilizations, these men—until recently, few were women—made new things intended to pass for the original but be less expensive, more abundant, or both. Some of the products of their genius or skill would, over time, come to be respected as honest materials in their own right. Some, such as celluloid combs, are today even collected as antiques, right beside their progenitors of horn and bone.

But often the imitators have been dogged by a kind of moral and ethical taint. They were not, obviously, dishonest men, these inventors and artisans; they were not forgers, counterfeiters, and confidence men; they'd never be hauled into court on charges of duplicity or fraud. And yet the materials they created are often viewed with such disdain beside the trusty, traditional, much-loved ones they've replaced that their handiwork bears a stain of falsehood and mediocrity. To some, indeed, it's all quite simple: The natural, original, and authentic are good. The synthetic, the imitation, and the artificial are tawdry and bad. End of story.

In a Depression-era article in the *Quarterly Journal of Economics* entitled "Economic Aspects of Adulteration and Imitation," a Stanford University scholar, Carl Alsberg, explored such terms as *imitation*, *artificial*, and *synthetic* as they applied to products like skim milk, reconstituted rubies, and imitation leather. They weren't the same, certainly, but had they anything in common? Did any general term, he asked, cover them all? In the end, he settled on "falsification"; it struck him as more apt than another candidate, "debasement." Neither term, of course, counts as flattering.

You can see the taint in Ian McEwan's novel *Atonement.* Paul Marshall is the white-suited visitor to an English country estate just before World War II; he is a chemist and thus, to one of his uppercrust hosts, "incomplete as a human being." While trying to land a large chocolate contract from the British army, Marshall is also working on an entirely new product. Seems he's "found a way of making chocolate out of sugar, chemicals, brown coloring and vegetable oil. And no cocoa butter." Chocolate, in other words, that is not chocolate.

Nothing good, it seems superfluous to say, comes of Paul Marshall.

Are the Paul Marshalls of the world, then, properly seen as debasers of civilization? As adulterators, shabby imitators, and soulless purveyors of the second rate? As those who would spread tackiness on everything they touch and lead us to a brave new world divorced from the natural and from nature itself? Or—to resort to an equally cartoonish vision—are they innovators, benefactors of humankind, dreamers and doers who bring utility and even beauty to millions who could not have the original for themselves because too expensive, too hard to come by, or both?

No out-and-out good guys or bad guys populate our story. But the borderland between the natural and the synthetic, the real and the faux, has become a battleground of our digital age, resident, for example, in the very phrase "virtual reality," a source of cultural discomfort. Everywhere today we find the copy, the imitation, the substitute: Audio systems promise high fidelity. Fidelity to what? To music made by live musicians, which they imperfectly reproduce. Color photocopiers forge currencies virtually indistinguishable from the original. Scientists clone genes, and sheep. Airline pilots train on simulators that imitate the sounds, sights, and lurches they'll encounter on real planes rising up through real oceans of air, obscured by real fog. Computer scientists equip us with virtual reality that, as time passes, seems ever less virtual, ever more real.

Is a generically modified tomato a real tomato? We don't *know* what's real anymore; we're not even quite sure what it means. This unsteady ground leaves many of us feeling troubled yet perversely fascinated.

Corfam, whose confrontation with leather we'll see played out in Chapter 7, wasn't the first faux leather, of course, and wouldn't be the last. As early as the 14th century, fabric was treated with special oils to make it resemble leather. In the 1870s came leatherette, much used for bookbinding. Around the same time, Japanese craftsmen, their names lost to us today, found ways to treat and decorate traditional Japanese paper so it could pass for the embossed Spanish leather that graced the halls of European palaces. In the early 20th century came Fabrikoid, which covered the seats of many a Model T Ford. In the 1950s came Naugahyde and the other vinyl fabrics, both derided and loved, a veritable icon of the faux. And in the years since Corfam, as we'll see, came many more, each a story of scientific, technical, or entrepreneurial innovation. Each borrowed from leather something of what made it leather. Each sought to transcend one or another of leather's limitations. Each compared itself to the real thing and pronounced it better. *Better than leather* was said of Corfam in 1963, but it, or something like it, could be heard in 2003 or 1883.

Today, leathers and fake leathers, both, are inescapable, part of our daily lived experience. The baseball gripped, then hurled, by the major league pitcher. The light suede jacket thrown over your shoulders on that first crisp day of fall. The book pulled down from a dusty library shelf. The clutch of bargain handbags in Filene's Basement. The thousand-dollar pair of exuberantly tooled Western boots. Your car's upholstery and its dashboard padding. The belt that holds up your pants, the gloves with which you grip your bike's handlebars. Your purse, your wallet, your appointment book, the trim on your canvas tote bag. Basketball and shoe, cell phone case and suitcase. The seat of your Breuer chair, the Nikes you wear in the backyard, the strapless sandals in which you dance all night. A confusion of textures and weights, but all bearing the telltale grain markings of leather. Some of them real leather, from the skin of cattle and pigs, alligators and deer. Others the distant product of oil piped from underground and transformed into polyurethane and polyester, polythis and polythat, that is blown and spun and squeezed and layered and pressed to look like leather. Some the cheapest, oiliest, creepiest, most odious stuff you can imagine, some a sensual heaven. Some real. Some faux.

Occasionally, these leathers and faux leathers well up into the foreground of our lives, forcing us to notice. By serving us, enduring through the years—the wallet, pulled from your pocket a thousand times, burnished with use. Or else by disappointing us—a seat cover ripping at the worst time, degrading, bleaching out, pulling apart. Mostly, though, we hurtle through life heedless of what we wear and touch, the materials we use a muddle, a blitz to the eyes and hands, ignored, falling away into insignificance against the larger pleasures and disappointments of over-busy lives.

But just this once, may we slow down a little?

That, anyway, is what I propose.

Faux Real is a book about leather and its imitators; it is about "natural" materials and the man-made ones that would replace them, and have replaced them.

It is about the people who bring new materials to the world and those who cherish one old material, leather, that's been part of the human experience for thousands of years.

It is about the surface sheen of things and what lies underneath.

It is about beautiful and beautifully made things and other things that are supposed to be but aren't.

It is about authenticity and imitation, counterfeits, copies, facsimiles, and fakes.

In September 2003 the American recording industry, stung by downloading of music and consequent loss of CD sales, tracked down, and brought suit against, 261 alleged perpetrators. The move spurred much debate, on the legality and ethics of copying in particular and on intellectual property more generally. "A Nation of Copiers," one writer, John Leland, branded Americans in the *New York Times.* Americans copied work off the Internet that was not their own, bought up bootleg Prada bags, read authors like Stephen E. Ambrose who'd cribbed material from other authors. "In a nation that flaunts its capacities to produce and consume, much of the culture's heat now lies with the ability to cut, paste, clip, sample, quote, recycle, customize and recirculate." Big brands like Tommy Hilfiger and Nike didn't bask in their reputation for quality but in their brand recognition, on the scant few millimeters on the surface of their shirts or shoes that bore a familiar logo or imprint. "In the brand market," wrote Leland, "value lies not in making things but in copying one's logo onto as many of them as possible."

That's what they've been doing to leather for hundreds of years now: Leather's "logo"—its distinctive look, its brand recognition, as it were, lies in its characteristic grain, the pattern of hair follicles, skin markings, and veins that together make it look leathery. And people have been embossing it on handbags, shoes, and soccer balls, on the covers of cheap bibles, the upholstery of Model Ts, the vinyl dashboards of Chevys and Mercedes for as long as anyone can remember. Take a women's dress that might otherwise slip into the vast faceless sea of others like it, sew in the right label, and you confer market value. And that's what makers of shoes, handbags, and car interiors have done by imprinting leather's surface grain. Sometimes the grain runs deep, sometimes unconvincingly shallow. But always, whenever the attempt is made, it's leather's logo they're after.

Actually, it's not just plain, undifferentiated "leather" they imitate but different *kinds* of leather, mimicking special treatments available from the leather tanner. For instance, one company offers a vinyl product it calls GLAZE that imitates "the elegant look and feel of polished antiqued leather, with prominent grain markings set against a glossy background." In another, SYMPHONY SUEDE, soft fabric floccing fills the low spots of an embossed pebble grain to suggest suede. VINTAGE imitates "waxy pull-up leather," another staple of the tanner's art, which forms distinctive streaks when stretched or creased. No matter that in some samples of GLAZE the embossing seems not to have taken, or that waxy VINTAGE isn't waxy.

Once, it was usually just surface grain that the imitators imitated, with scant effort to replicate leather's other qualities. But with Corfam, certainly, and in recent years others, it's not just the look but something like its leathery essence they've gone after. Getting it to breathe. Hinting at its weight and substance. Giving it, on its underside, that filamentary abundance that, indeed, resembles once-living flesh.

Leather and its imitators can be seen as something like mistrustful brothers. Being siblings, they bear a family resemblance; they are, after all, *supposed* to bear a resemblance. But one, the imitator, has traditionally borne a stigma as the family's black sheep—seen as never quite good enough, forever trying to play catch-up with his big older brother, the first and original.

Now, though, it can seem that some younger brothers have begun to stand up straighter and prouder. You'd hardly have believed it possible, but it's as if Naugahyde's children have grown up, taken charm school lessons, learned how to dress and talk; they don't offend so much anymore with spurious smell or oily touch, don't wear and tear prematurely, but can hang out in the best society, say all the right things, not embarrass themselves, come up to the standards of their classy, chisel-jawed older brother. So in the end leather and some of its imitators are closer to identical twins than mere siblings; you need a microscope to tell them apart.

In *Faux Real*, I propose to place these common materials—leather in its many permutations and faux leather in *its*—under another sort of lens. To stop, slow down, and see where each came from, how they landed on our hands, covering our feet, and slung over our shoulders; look at the science and technology behind them and at how they're made; consider their qualities, strengths and shortcomings, the vagaries and peculiarities of their use; their place in our lives.

In this one story we will see tensions between authentic and imitative, natural and synthetic, real and faux, that run deep through modern life. These tensions can be felt in the shoes we wear, yes, but also in the food we eat, the ways in which we work and play, in the things we treasure and those we throw away. Our old world of wood, stone, cotton, and leather has been metamorphosing into one of artificial sweeteners, Formica tabletops, and computer pixels, these becoming both more ubiquitous and more taken for granted, the rush toward the New Things seeming ever more headlong and inescapable.

As we'll see, the quest to create materials that mimic leather grants us a sharp, surprisingly versatile instrument with which to probe these trends of many centuries' standing and lets us explore eternal dichotomies that run like a river of uncertainty through our culture: the pull on us of the natural, the authentic, and the lushly real but also the power and appeal of all that's frankly fake and, always, the slippery, porous boundaries between them.

I do not assert that in the story of leather and its imitators lies some hidden key to understanding modern civilization. I make here a more modest claim, one more akin to the novelist's, who finds in a rural Mississippi county all the foibles of humanity, or the biologist's, who discerns in a thimbleful of pond water all of nature: We can chart the departure from our lives of traditional materials—"authentic," "natural," "real" materials—by charting the sustained human effort, over the past few hundred years, to imitate, supplant, and synthesize one of them.

Why leather?

Because I like leather, always have. Best to 'fess up once and for all.

From my childhood I remember my mother's purse, with its slim, Mondrian-like strips of delicate leather, its forest greens, its beiges and browns, its smoothness and sheen. I remember my father's leather slide rule case, hard and firm to protect the mahogany instrument within, its worn burnt-orange surface discolored by chemical stains that reached through to the skin beneath, and how, in the fine tangles of fiber on its underside, you could see it really *was* skin. I've made things of leather, including briefcases, pocketbooks, sandals, and shoes. They lacked much by way of craftsmanship. But the weight of the hide in my arms, the way it cut under the knife, the feel of it under my fingers—these all sustained me against the evidence of my own want of skill, and left me with a kind of pleasure in the material itself.

So I like leather, yes, and I don't think that to declare such affection for a particular material is so unusual. Cabinetmakers love fine woods, potters delicate white porcelain. Polished ebony, lustrous gold. Real materials. Authentic materials. Materials that have been with us for virtually all of human history, that delight the eye, reward the touch, stir affection. Oak. Brick. Silver. Silk. Is there a word for it, this love of particular materials? *Materialophilia?* Do synthetic materials generate affection like that? Does Formica? Vinyl siding? Naugahyde? Asphalt? Is there some primeval hold that the old materials still exert, even as most of our lives are spent with the new?

"I like your coat," says a young man to the Sandra Bullock character in the film *28 Days*, eyeing her attentively.

"Thank you," she replies with downcast eyes.

"Is that leather?"

"Yes."

"Not vinyl?"

"Nope."

"You believe in killing animals?"

"Yes."

"For clothing?"

"Absolutely."

"So do I."

She's the real thing, the leather of her coat a talisman of authenticity.

Among many today, 1950s sensibilities stand reversed. A resurgent American crafts movement—jewelers, quilters, potters, musical instrument makers, leathercrafters, woodworkers—abjures any but fine materials finely worked. Well-off consumers covet them. The old things—the natural, the authentic, the original—possess cachet, command high prices: Oak cabinetry. Pure wool. All cotton. Genuine leather.

In some ways more than other natural and traditional materials, leather *stands for* the authentic and the genuine; Genuineleather, like a single German word, is how we think of it. Its animal roots etched in its pores and in the swirls of its grain, leather serves as cultural shorthand for the virtues of the real over the synthetic, the original over the copy, the genuine over the fake, the luxurious over the shoddy and second-rate. It thus represents a pole from which modern life, clogged as it is with copies and fakes, sometimes seems ever to be receding. In leather's fate at the hands of its imitators we may track the transformation of our material world.

2

"Let the Good Work Go On"

Little boxes.

Once they held engraved calling cards, ladies' fans, gold tie studs. Shelly Foote, curator at the Smithsonian Institution in Washington, D.C., presents them to me in a shallow cardboard tray pulled from the Smithsonian's vaults. As I open their lids, fluffy satin or velvet cushions rise up to meet my touch.

One box, covered in sumptuous, butterscotch-toned leather, from a shop in Rome, has a clasp that even after most of a century opens and shuts with a sure little snap. Another, from the 1890s, bears the imprint of a shop on boulevard des Capucines, Paris; it's a fan box, sheathed in maroon leather, once owned by the daughter of American statesman Elihu Root, donated to the Smithsonian after her death. There are about 20 of them in all—oval, round, rectangular, and clamshell shaped, in different colors. But viewed from a few feet away, they look more alike than unlike, as if they belong together here, all dainty, precious, swaddled in leather.

But for some the leather is real only where it shows, fake on the sides or bottoms; the contrast reminds me of the fine facades and scraggly back alleys of the row-house neighborhoods of Baltimore where I used to live—the fronts, with their fine, imposing public faces, the alleys, for the trash, that visitors don't see. In a Tiffany case, *circa* 1895, I see where these two worlds, real and fake, meet—the border where one material is glued down beside the other. On one side of it, the old abraded leather shows wisps of fine, fibrous blue; on the other, clouds of white dots, where the pebbly leatherette's thin coating has worn through, reveal the paper underneath.

Today, "leatherette" can mean about any leatherlike material you like—vinyl, maybe, or coated paper, or coated fabric; bibles are bound in it, chairs upholstered, and all it tells you, thanks to that dismissive suffix *-ette*, is that it's not the real thing. But once, in 1875 England, Leatherette—make that an uppercase L, if you please—was someone's proud baby, a material its London representatives defended, and whose virtues they bruited about, in a small gold-embossed volume. The book bore *Leatherette* as its title, promoted Leatherette, was bound in Leatherette. And it contained swatches of Leatherette that today seem as intensely crimson, purple, and plum, as when first glued down onto the page more than a century ago.

The maker of Leatherette was Harrington & Co., which billed it as "a substitute for the commoner sorts of leather." The stuff sold for 42 shillings per 100 "skins," an eighth the price of the best real skins, and could be used to cover opera glasses, line purses and jewelry boxes, panel walls. A varnished, pressed, long-fibered felt, it bore a surface grain drawn from a real leather hide. And it was dyed all the way through, as Harrington's took care to note: no papery white patches showing up at the first signs of wear, the way that Tiffany's case did. Worn corners of this 130-year-old volume show off color so vivid and deep it seems downright gaudy.

In the spring of 1875, Harrington's deluged British trade journals with samples of its new product and must have been delighted with the laudatory reviews it inspired. One said Leatherette was "stronger than any sort of leather except the very best, and possesses the special advantage of being soft and warm to the touch like leather." A bookbinder admitted that his own workmen displayed stubborn prejudice in favor of leather and against its imitators, but he wished to pay "humble tribute to the energy and spirit of the inventor who has brought 'Leatherette' to its present perfection." Another correspondent wrote that a near twin of Leatherette, called Powell's Pelt, was "probably the closest imitation in appearance to a natural product that has ever been obtained, and seems to be so perfect a facsimile as to court microscopical examination." Samples of it had "deceived connoisseurs in the leather trade."

But that fall, in *The Bookseller,* came an attack, penned by "A Practical Bookbinder" of three decades' standing. Leatherette, he wrote, showed an "entire absence of the suppleness and elasticity of leather." This made it "worthless," giving proof "of the well known adage, that there is nothing like leather." Two issues later, he was back, now mounting a far broader attack. The Leatherette-bound pocket diaries he'd inspected struck him as "cheap and nasty in appearance, unpleasant to handle," neither soft nor supple. Besides, if Leatherette had any intrinsic merit, "why is it necessary to trick it up to make it imitate leather so closely?" Why? To deceive the public, *that's* why. "The article is a sham" and deserved to be exposed as such.

Harrington's shot back: "There can be no sham where no deception is intended." None, it claimed, was. Yet testimonials in its own little book suggested otherwise: Leatherette so nearly resembled fine leather "that none but the initiated would be able to distinguish the points of difference." And: "It is so good an imitation that a superficial observer would, in nine cases out of ten . . . take it for the real article."

Played out in sallies back and forth in these old British trade journals were themes that would crowd in upon imitation leather—and imitation-everything-else, for that matter—in all the years since. Technical innovation versus tradition. Lower cost set against lamentations of inferior quality. Benign emulation countered by charges of deception and sham.

Leatherette, patented in 1872, was not the first imitation leather. Indeed, appearing only about midway through the great age of Victoria, with the industrial revolution at its giddy height and inventors, engineers, artisans, and entrepreneurs caught up in a fever of making and selling, it invited comparison not only to genuine leather but, as one noted, to "its numerous imitations."

The imitations had blossomed with the coming of rubber, a material whose history would curiously parallel that of leather even to the present day. Obsessed with finding a way to treat natural rubber so it wouldn't soften in the sun or crack in the cold, Charles Goodyear in 1839 finally found it. "Vulcanization" was the name a British competitor gave the process, which relied on sulfur and heat, in just the right proportions, at just the right times, to make it work. The subsequent growth of the rubber industry brought waves of new products, on both sides of the Atlantic. One was imitation leather.

Or rather, imitation leathers, plural; countless patents were issued over the next 75 years for artificial leathers using rubber—or *caoutchouc*, its South American Indian name—as a key ingredient. Combined with oils, pigments, cloth, cork, scrap leather, and other materials, and duly crushed, mixed, ground, kneaded, melted, or dissolved, these rubbery concoctions could be rolled into sheets or applied to cloth and made into something reminiscent of leather. One account from 1854 tells of

> a most perfect imitation of morocco [leather] by the application of a preparation of caoutchouc, or gutta percha, to the surface of a plain woven or twill cotton cloth. The surface is corrugated in imitation of morocco, and is coloured and varnished so as to present all the external appearance of that kind of leather. The elasticity is perfect, showing no tendency to crack.

In early accounts of faux, quasi, and ersatz leathers, of course, their elasticity and other qualities were *always* perfect. The imitation *always* looked just like the real thing—or better. From the early 19th century comes a claim of table covers of painted and varnished canvas that would "probably answer better than leather." An 1847 history of shoes and boots tells of a material pitched to tenderfooted women, known as pannuscorium, or leather cloth, which had proven popular among that "class whose feet require something softer even than the softest leather." Most of these artificial leathers were nothing more than cloth of cotton or jute with a coating that, once dried or cooled, was more or less leathery and could take a leather grain. But within this broad theme, the permutations were endless.

Inventors used materials most of which were little removed from their plant, animal, and mineral origins. Gutta percha was a yellow latex derived from trees in the South Pacific or South America used for products as various as chewing gum, insulation, and golf balls. China clay, or kaolin, named for a hill in China where it was mined for centuries, was a white powdery clay that, when mixed with water, became moldable or, with more water, formed a slurry often used in papermaking. There was gelatin, and resins, and oils, and butter, and wax, and straw, and wood fiber enough to befit a witch's brew; all of them were sometime constituents of early imitation leathers.

One prominent early manufacturer was Storey Brothers, a Lancaster, England, firm that by about 1850 was producing a "leathercloth" used for lining the coaches of a small English railway and soon was being used for horse-drawn carriages as well. By 1862, Storey's was selling the stuff in three qualities and a dozen colors, using several different kinds of cloth. In time it was sold in India, where it was used to make rickshaw covers.

Later, the company had machinery to do the job, but in the early days—this account probably dates to the 1860s—it was all done by hand. Across a simple wooden platform about a dozen yards long workers stretched a length of cloth, then troweled onto it various rubbers, oils, and pigments. One version of this generic technology, destined for flooring, became known as linoleum. A second was oilcloth, which replaced cotton or linen on some tables; these coverings were typically printed with designs, some of them quite intricate, and given grandly inflated names, like MADAGASCAR DAMASK or VERSAILLES TAPESTRY.

A third was "American Leathercloth." Here, they weren't *decorating*, weren't printing pretty pictures of Don Quixote or Dutch gardens; the aim was to imitate another material entirely. The prime target was leather's surface grain, which you'd try to copy onto a small, hand-cut embossing tool, maybe 6 inches long, crowded with pits, veining, and furrows, whose detail then had to be transferred to a large roller several feet long. The embossed leathercloth was used in bookbinding, baby carriages, furniture, and apparently even boots and shoes—all the customary markets for real leather.

So: Leatherette and leathercloth. . . .

There was leatherine, too, with a textile base, its coating impregnated with rubber.

And leatherboard, pulped fiber and chalk glazed with a mixture of starch, gelatin, and turpentine.

And various sorts of scraped, pulped, or powdered leather scrap, compounded with agglutinants like rubber, various resins, or boiled linseed oil.

And fibroleum, from animal glue mixed with paper pulp and glycerol, then pressed. Its inventor claimed for it "an elasticity as lasting as original leather."

None of them were really much good.

This is not fair, of course, and not strictly true. Amid the rough and tumble of late 19th-century commerce, some of them found markets, made money, and resembled real leather closely enough to suit their users. And yet with each of them there was always something wrong. Recall that Leatherette's persistent critic lamented its lack of suppleness; maybe it looked like leather, but it didn't feel like it. Some of the early paper-based leathers fell to pieces when they got wet. To the coating of another would-be faux leather they added castor oil to improve its flexibility, but the result, as one account had it, simply did "not bear any resemblance to real leather." As for Storey's linseed oil-based leathercloths, their use in hot, humid places like India pointed up their deficiencies. "Occasionally," a company biography admitted, "sad examples were found of what could happen—stock kept for some time would emerge in a solid and disgusting lump, of no further use to man or beast."

If the faux leather of the age wasn't very good, it was not for want of trying. The late 19th and early 20th centuries were an era of spirited commercial and technological enterprise—a second industrial revolution of steel, chemicals, and machines, following that first revolution borne on the shoulders of steam power. Factories were going up. Great industrial combinations were at work. It was the age of Goodyear and Edison and Westinghouse and the other iconic inventors, along with thousands of others less well known today. And much of what these inventors did was imitate. Imitate what? Everything.

In her acerbic 1913 novel *The Custom of the Country*, Edith Wharton's social-climbing character, Undine Spragg, comes away disappointed by dinner at the Fairfords' shabby house, in part because

> instead of a gas-log, or a polished grate with electric bulbs behind ruby glass, there was an old-fashioned wood fire, like pictures of "Back to the farm for Christmas"; and when the logs fell forward Mrs. Fairford or her brother had to jump up to push them in place, and the ashes scattered over the hearth untidily.

The period of Wharton's novel, and before that the whole Victorian era, saw substitute materials made possible by new industrial processes imitating and supplanting nature at every turn—even the glow of fire. Imitation was in the Victorian air.

In *Cheap, Quick, and Easy*, Pamela H. Simpson, an art historian now at Washington and Lee University, writes of the imitative architectural materials popular then. Pressed metal ceilings, concrete blocks, embossed wall coverings, and linoleum were replacing engraved wood, sculpted granite, fine leather. It was as if humanity, just then, gloried in a cult of imitation. As American studies historian Jeffrey Meikle has summed up the era, imitation embodied

> a brash spirit of technological exuberance. . . . Building fronts and cast iron assumed outlines and textures of brick and stone. Automatic machinery produced wood carvings the intricacies of which seemed a product of painstaking craftsmanship. Stamping and moulding of papier mache liberated ornate carved effects in furniture and nicknacks from any reliance on wood at all.

Of course, critics caviled. For many, sneers and lifted eyebrows were the only fit response to what they deemed the ugly, cheap, derivative gimcrackery infiltrating the Victorian landscape. "Trick and falsehood" might fool men, but never "the all-searching eye of God," declaimed Augustus Welby North Pugin, a noted Gothic revival architect who by the early 1840s saw in cast iron, artificial stone, plaster, and the graining of materials to resemble wood the sorriest technological sham.

This last practice inflamed the seminally influential art critic John Ruskin, who in books and countless essays campaigned against industry's usurpation of human effort, intelligence, and talent. Of graining, Ruskin wrote this:

> There is not a meaner occupation for the human mind than the imitation of the stains and striae of marble and wood. . . . I know not anything so humiliating as to see a human being, with arms and limbs complete, and apparently a head, and assuredly a soul, yet into the hands of which when you have put a brush and pallet, it cannot do anything with them but imitate a piece of wood.

Moral fervor runs through Ruskin's response. And certainly there was much ought-and-should among critics of the new imitative technologies. The English poet, artist, and designer William Morris asserted, straightforwardly enough, that those designing for machine production ought not "make stone look like ironwork, or wood like silk, or pottery like stone." Period.

But that's what British and American industry continued to do. As Pamela Simpson began studying imitative materials, she wondered this: With critics so quick to dismiss imitation as "vulgar, cheap, and tasteless, why were the materials so abundant in the built environment? Was it simply bad taste?" No, she decided, they benefited people: "Machines made ornament available to the masses." Lament bad taste all you like, but poor homes once bare, unornamented, and unfinished could now boast a spot of color, or a hint of flavorful design, to enrich lives otherwise deprived of them.

Imitation was everywhere. And, viewed through the right lens, it was good.

On April 7, 1884, two men affixed their signatures to a British patent application for "Improvements in the Production of Compounds Containing Nitrocellulose Suitable as Varnishes and for Making Leather Cloth." The two were Joseph Storey, the Lancaster maker of linseed oil-based leathercloth, and William Virgo Wilson, described as a "colour manufacturer." In the patent awarded seven months later, the two specified that nitrocellulose could be dissolved in amyl acetate to produce a liquid with "the consistency of varnish" or, more thickly, as a paste or dough. Mixed with pigments and oils, particularly castor oil, spread over cloth, and then let dry, it could then be used for a number of products, including artificial leather.

Amyl acetate and castor oil were the new kids on the block—the new ingredients, the improvement that formed the basis for the patent in the first place. But it was nitrocellulose that was the object of everyone's attention, nitrocellulose whose properties, qualities, and industrial reach everyone wanted to expand. Nitrocellulose came from cellulose. And cellulose came from wood—and, for that matter, most everything else in the plant kingdom; it is the most abundant of all naturally occurring organic substances. Invent something useful made from cellulose and you had a cheap, virtually infinite supply of the raw material at your disposal.

Christian Friedrich Schonbein was a German chemist, a 45-year-old professor at the University of Basel in Switzerland. In the fall of 1845, he was working with mixtures of sulfuric and nitric acids and took to studying their effect on organic substances like paper and sugar. And cotton. Dip a swab of cotton in his mixture and it came out looking pretty much unchanged, except that now it burned, with great ferocity and almost instantaneously. Might this new material be put to military use? The following spring Schonbein advised the authorities in Basel, who apparently experimented with it. He predicted that what he called "explosive cotton wool"—which became better known as guncotton and chemically was nitrocellulose—would ultimately replace the various gunpowders then in use. It did.

The first large English factory to make guncotton exploded in 1847, killing all hands. The following year 1,600 kilograms of the stuff blew up in Bouchet, France, devastating the surrounding countryside. *This* was nitrocellulose. Yet so was the material figuring prominently in the Wilson and Storey patent. How could it have a future as the basis for artificial leather?

In fact, nitrocellulose was something of a misnomer; it was never just one thing. The product of a reaction between cellulose and nitric acid—the sulfuric acid in Schonbein's experiment served as catalyst and didn't figure directly in the reaction—nitrocellulose is today called cellulose nitrate; two names for the same thing. In 1911, chemist Edward Chauncey Worden published a two-volume, 1,200-page tome, *Nitrocellulose Industry*, that included a chapter titled "Gun Cotton, Smokeless Powder and Explosive Cellulose Nitrates."

Many cellulose nitrates.

Cellulose itself is a natural polymer built up from a basic repeating unit. That unit is $C_6H_{10}O_5$, where C is carbon, H hydrogen, and O oxygen; the subscript 10, say, indicates there are 10 hydrogen atoms in each unit. Add nitric acid, HNO_3, to cellulose, and a nitro group, NO_2, supplants a hydrogen atom in the cellulose subunit. That yields cellulose nitrate. Its chemical formula, $C_6H_9(NO_2)O_5$, straightforwardly records the loss of a hydrogen and its replacement by a nitro group. But as it happens, this is only crudely—and insufficiently—correct. Because why should only one of those 10 hydrogen atoms in the cellulose yield to a nitro group? Why not two, three, many?

In fact, hydrogens can be supplanted by nitros to a greater or lesser extent, depending on the controls placed on the reaction, such as temperature, acid concentration, and so on. So variant cellulose nitrates are possible, each distinguished by its "degree of nitration," or how many hydrogens give way to nitro groups. Each can be represented by its own name and formula; dinitrocellulose, for example, with two hydrogens replaced, would be $C_6H_8(NO_2)_2O_5$.

The properties of each nitrocellulose depend with exquisite sensitivity on its degree of nitration. Any of them looks pretty much like any other or, for that matter, the raw cotton from which it's made. But its solubility in various solvents, for example, and of course its explosibility differ profoundly; you can wind up with Mr. Hyde or Dr. Jekyll. In a sample that tests out at about 13 percent nitrogen by weight, for example, you've got an explosive that can blow up the lab. But limit nitration to $11^1/_2$ or 12 percent nitrogen and you've got a material that for years went into motion picture film and the handles of ladies' hairbrushes. Dissolve this less nitrated nitrocellulose in alcohol and ether and you wind up with what by the mid-19th century had its own name—pyroxilin.

When dried, pyroxilin—pronounced py-ROX-i-lin—formed a hard, translucent coating that during the American Civil War was used as a wound dressing. A little later an English chemist and inventor, Alexander Parkes, mixed it with camphor, a white, waxy substance derived from the wood of an Asian tree, and wound up with celluloid, most familiar today as the material in Ping-Pong balls. Celluloid was "the pioneer plastic," as historian Robert Friedel called it in his book of that name, and for more than half a century it was the basis of a great industry. Refined through numerous patents, by dozens of chemists and entrepreneurs, and carrying a slew of trademarked names, celluloid became motion picture film, collars, piano keys, billiard balls, and combs, many of them worked to resemble ivory, horn, or wood.

Early on, some had tried using pyroxilin to coat cloth for imitation leather. But *hard* and *translucent,* its most notable qualities, won't remind you of leather. So as Worden wrote in 1911, early formulations "lacked that pliability which is present in so noticeable a degree in leather, and the absence of which prevented serious competition of the artificial with the real skin." But then in 1884 came Storey and Wilson, with their new twist, the castor oil and amyl acetate recipe, and this did indeed mean serious competition for leather.

Amyl acetate, volatile and flammable, looked about like any other colorless liquid. But by taste and fragrance it was utterly distinctive:

bananas. Today, it's used as a banana flavorant and, in fact, is found in natural banana oil. As for castor oil, it was the same thick, notoriously foul-tasting liquid, derived from beans grown on fast-growing plants in India and Brazil, famously prescribed as a purgative by cold-eyed physicians and parents. By the 1884 patent, a coating made from these two materials, along with pyroxilin and suitable pigments, could be applied to cloth and embossed with a leathery grain, making for a new, presumably superior artificial leather; it covered at least one little box in Shelly Foote's collection at the Smithsonian.

Nitrocellulose-based leathercloth was made into book bindings, automobile seats, desk blotters, and baby carriages and in its day was about as ubiquitous as Naugahyde is in ours. In England, Storey's marketed its version under the name Queen's Leathers. Competitors made nearly identical products under the names Rexine and Pegamoid; these became household words in Europe, and *pegamoide* still means artificial leather in Italy.

In America several companies made versions of the material, each with its own balancing act of solvents, oils, and pigments, and each with its own trade-off between durability and suppleness. There was the Tannette Manufacturing Company, established in Springfield, New Jersey, in 1890; it later became the Boston Artificial Leather Company, which made Moroccoline, a reference to the soft morocco leathers used in bookbinding. Other companies came and went and with them other products, all variations on the same theme. One of them was Keratol, another Marokene, a third Texaderm. You could make a poem of their names.

A fourth company, established in 1896 in Hohokus, New Jersey, was the American Pegamoid Company, which became the New York Leather & Paint Company in 1898. In 1900 a group of businessmen raised $5,000—perhaps $150,000 in today's money—to lure New York Leather to Newburgh, New York, on the west bank of the Hudson River 50 miles north of New York City. Two years later the company, now known as the Fabrikoid Company, moved to the site of an abandoned felt-making factory on the edge of town. By 1908 it was annually shipping almost 2 million yards of artificial leather for use in shoe linings, carriage trim, automobile tops, baseballs, and much else. By some measures it was the largest manufacturer of artificial leather in the world.

About this time the Du Pont Company was scouting around for potential acquisitions. More than half a century before it introduced Corfam,

Du Pont bore little resemblance to the company it would become; it was big, successful, but not a "chemical company." Since its founding in 1802 by a refugee from the French Revolution, it had been making explosives and almost nothing else; visit today the Hagley Museum in Wilmington, Delaware, and the carefully preserved stone ruins of explosives-making buildings along the Brandywine River make for an idyllic tableau. But while Du Pont explosives were used benignly to build tunnels and clear underwater navigational obstacles, the company's fortunes inevitably rested on military uses. Among its products was smokeless powder, made from nitrocellulose, a business threatened by persistent grumbling in Washington against "The Powder Trust," the explosives monopoly led by Du Pont; perhaps the government would start making explosives on its own.

Facing the specter of excess manufacturing capacity, Du Pont began looking into what else it might do with nitrocellulose. In December 1908 a detailed report on "Artificial Leather and Allied Products" was submitted to the company's executive committee. It surveyed other companies in the business, some of which were already customers for Du Pont nitrocellulose. It reviewed the technical problems of making artificial leather. It came up with estimates for the investment needed, down to toting up boiler houses and embossing plates.

In the end, Du Pont didn't start from scratch in the business but bought into it. In its first step toward becoming the conglomerate it is today, it paid a little more than a million dollars for the Fabrikoid Company. Soon it was making thousands of yards of artificial leather each day.

Du Pont's Newburgh factory was a ragtag collection of low, modestly scaled structures, some of them not much more than tin-roofed shacks, one of them an old stone building said to predate the American Revolution. They flanked a creek running down the middle of the site spanned by pipes, dams, and wooden footbridges. The big coating rooms stood toward the middle of the property. To the east, on the far side of the creek, were buildings that stored nitrocellulose or various solvents or that were used to mix the coating.

First, they'd dye the cloth to match the coating mixture, then squeeze it between rollers to remove excess dye, dry it, and wind it back onto a roll, maybe 300 yards of it at a time. Then it was on to the coating room, where they'd apply the first of several layers of the magic mix; a typical formula called for 40 pounds of cellulose nitrate to 55 of castor oil, plus 8 gallons of amyl acetate, other solvents, and pigment enough to get you the right shade. The final consistency was that of molasses or jelly.

The fabric to be coated passed beneath a stationary steel blade called a "doctor" knife, an old term in the mechanical arts for a blade that removes defects, adjusts, or regulates. Here, it served as a kind of dam for the trough of jelly above and behind it, letting an adjustably thin layer of the stuff seep under it and onto the surface of the slowly moving fabric. The first, or skin, coat was usually heavier on castor oil, to get something like the needed suppleness; later coats were heavier on cellulose nitrate.

The solvents were made to evaporate in large ovens, heated to about 185 degrees Fahrenheit. Roaring fans carried off the vapors for recovery. Once dried, the surface was normally a little pitted, so now the coated fabric was "calendered"; the word comes from the same Latin root that gives us cylinder and means to smooth or glaze between hot rotating rollers.

Finally, the newly pristine surface was embossed using steam-heated rollers or flat-bed presses that bore the imprint of a particular leather grain; the flat beds were more expensive but gave better definition, successive lengths of coated cloth being fed between the jaws of the hot plates, pressed, and advanced. "The fidelity with which leather grains are reproduced," a Du Pont publication noted in 1919, "has curiously, but very naturally, resulted in giving leather substitutes a much more beautiful grain than the average grain of real leather. . . . What was a case of isolated perfection in real leather becomes the standard in Fabrikoid." By the late 1920s, Du Pont maintained more than 100 distinct leather grains, which were applied to coatings that came in thousands of colors and shades.

Over the years the process was refined and the chemistry tweaked, but this straightforward technology, in this unassuming factory, basically got you Fabrikoid artificial leather—6 million yards of it in 1915. It was already more than a quarter century since the first Storey-Wilson patent and Fabrikoid technology never exactly ranked as cutting edge. As one 1923 account of the Newburgh operation conceded, "The operations of coating, embossing, and finishing can hardly be called sciences"; they were the work of artisans and operatives more than of chemists and engineers. Making Fabrikoid was fraught with problems—duplicating colors, recovering solvents—but they were routine problems.

The more persistent frustration, especially during the early years of Du Pont's reign, was getting Fabrikoid to look, feel, and function more like real leather. In 1910 the split leather that was its chief competition went for about 28 cents a square foot, while Fabrikoid cost 17 cents; manufacturers ordered Fabrikoid for their products because they didn't want to spring for even the cheapest grades of inferior leather, hoping that the armchairs, book covers, or automobile upholstery they made would be close enough to the real thing to suit their customers. It was achieving this *close enough* that worried and troubled Du Pont engineers and technicians.

On December 28, 1915, Fabrikoid salesmen gathered at the Elks Club in Newburgh for the fifth annual dinner of the Du Pont Fabrikoid Knockers Club. Over turkey and blue fish, creamed onions and Prussian potatoes, the evening capped by dessert and cigars, they had much to celebrate as they looked to the new year; shipments were up by more than half over 1914. We don't know today whether they belted out the Fabrikoid Yell or, if they did, with how much enthusiasm, hilarity, or abandon. But it was printed in the program, and it is not hard to set our minds back to that early winter evening almost a century ago and imagine:

What do you think of business?
Fine – Fine – Fine!
What do you think of your bonus?
Very – good – Eddie.
What will you say if they ever take it away?
Poor – Paul-ine
Hula – Hula – Hula
'Pon – my – word
Better than the real hide.
Fab – ri – koid.

But Fabrikoid was *not* "better than the real hide," never even as good, and Du Pont executives, chemists, and engineers knew it.

"I have been disturbed recently that we may be led into either a definite or implied guarantee that Fabrikoid will stand 'as well as leather,' or better than real leather, for use in automobiles." The date was May 14, 1913. The letter's author was 37-year-old Irénée du Pont, an 1897 MIT graduate, one of three du Pont cousins who over the next 40 years would preside over the company's expansion. He was writing to William Coyne, the company's sales director.

Du Pont was doing well with Fabrikoid, selling scads of it to the automobile makers, especially Ford. "I saw six finished cars come through inside of two minutes," a Fabrikoid sales rep reported after touring a Ford assembly line. And almost two-thirds of the automaker's huge output bore Fabrikoid tops and upholstery. Ford was apparently little inclined to revert to real leather. But Irénée du Pont was worried: They were all but assuring Ford and the others that Fabrikoid was as good as leather, and it wasn't.

The Du Pont name was at stake. Two years earlier Coyne had claimed in an internal memo that Fabrikoid was actually better than many kinds of real leather, but reported resistance to its use among automobile companies worried "it would cheapen their cars in the eyes of the public." Du Pont had largely overcome their resistance but now risked being found liable by Ford or other manufacturers: Fabrikoid was good, especially for its price. But, wrote Irénée, it "apparently is not as good as leather and it should be sold with a clear understanding that . . . [it is] not an improvement or the equal of leather." In a handbook going to its salesmen a few years later, the company likewise conceded as much: Genuine leather was softer, warmer, more satisfying to the touch—"more beautiful than Fabrikoid by that narrow margin beyond which science cannot go in imitating nature." Could they eliminate or reduce that margin?

In September 1910, just two months after the Fabrikoid went through, chemist H. F. Brown wrote Irénée du Pont urging "development of a material capable of giving to artificial leather the odor of real leather." Fabrikoid, it seems, stank. Even at its best, by one report, the solvents used in its manufacture gave it "a characteristic and frequently objectionable odor." And Fabrikoid *wasn't* always at its best; sometimes the castor oil turned rancid. That's not what you wanted as you took your lady friend for a ride in your Model T or packed your new luggage for a trip. Needed, said a report coming out of the Experimental Station in November 1911, was a leather smell that might blot out the odor of chemical solvent or of castor oil turned bad.

In those days its personnel numbered fewer than 70, yet even then the Experimental Station was *big*—big in concept, well on its way to becoming a model for industrial research, alive with energy and imagination.

And now its chemists set out to develop an artificial leather that would smell more like the real thing. They tried oils drawn from the destructive distillation of wood, such as "oil of Russia leather," which bore a hint of tar sweetened with what the experimenters thought might be lavender. They tried oils of ginger and hemlock, cypress and patchouli. They gathered extracts from solutions used at tanneries and from hide shavings.

They mixed the products of their experiments with regular pyroxilin jelly, smeared it on cloth samples with a spatula, trimmed them into 1-square-foot samples, labeled them, laid them out to dry, inspected them a week later, and recorded the results—which were never very good. "A very disagreeable odor remained," was recorded for one group of samples. From another sample, they did get a leatherlike smell, courtesy of an extract made from leather scrap, but it apparently never made it into production.

Smell was one way Fabrikoid didn't stack up to real leather. In coming years, Du Pont researchers confronted many others. Irénée, as assistant to the general manager, would sometimes pass down technical suggestions from elder brother Pierre, also MIT-trained. Maybe Fabrikoid wouldn't peel so much if the cloth were impregnated with gelatin, treated to make it more like what "lies between the fibres in real leather." Something needed to be done about how Fabrikoid stretched along the diagonal, damaging the embossed grain and emphasizing the material's "clothiness." And on exposed corners the coating wore too readily. The result was puffy little protuberances of underlying fabric, wear spots embarrassingly obvious against the shiny finish; these, a later report put it, exposed "the artificial nature of the product." Such failings, as Irénée summed up his brother Pierre's suggestion in a December 1914 memo, argued for research toward a new Fabrikoid that might "measure up with the best quality of leather."

Needed was something new, something beyond the old pyroxilin-castor oil orthodoxy. "Mr. Du Pont," wrote Frank Kniffen, head of the company's Fabrikoid subsidiary, after meeting with Pierre and Irénée the following May, "suggested the careful study of real leather by microscope, strength and elasticity tests, etc., as a guide."

The first result of these proddings came in a report issued in June 1915 that reads like an exercise in pure imagination. It told of no experiments, cited no studies. It was as if the order had come down to simply read, think, indulge in any wayward vision you like, and spew out ideas—

without, just yet, having to subject them to test. The report, "New Types of Artificial Leather," was essentially a roster of ideas, 83 of them, drawn from German, British, French, and American patents, culled from technical articles, or initiated by Experimental Station chemists, all broken down by category.

The first group consisted of variations-on-a-Fabrikoid theme—a woven fabric, coated; for example, maybe impregnate the fabric, not just coat it, with rubber, celluloid, or linoleum cement, thus making the woven structure invisible. The second approach was to abandon the regular fabric backing and try something else, like felt. Or maybe two or more plies of ordinary fabric whose weaves ran in different directions. Or specially prepared paper. Or start out with cotton or rags as raw material, chemically treat them, then impregnate a felted fabric with the result. The final broad strategy was to press together various fillers and binders. You could blend animal hair and raw hide with sugar and glycerin. Or mix rubber with vulcanized fiber, chalk, asbestos, various chemicals, and vulcanize the whole mess. Experimental Station chemists cited one French patent—it reads, naturally, like a recipe—where you "mix casein or albumin with alkali solution until it gets a syrupy consistency. Stir in tannin. Add a caoutchouc solution in oil, then a mixture of linseed oil and sulphur. . . ."

If they weren't enjoying themselves hatching these ideas, each more unlikely, sometimes downright sillier, than the last, then these folks just didn't know how to have fun. Over the next few years a handful of ideas were pursued, most dropped, others taken up in their stead. From the Experimental Station came a succession of studies, experiments, literature reviews, field tests, and patent searches. Test pieces were made, special fabrics ordered, coatings applied, mixtures formulated, results tabulated. Pierre du Pont egged them on. "Let the good work go on," he replied to one report. "There is encouragement in what you have done."

But it never came to much. The cover letter to the report seen by Pierre noted that station technicians worried mostly about clothiness and diagonal stretch, scant attention being given to other prime qualities of leather, "especially softness and pliability." So, it added, "we are submitting this report and samples without in any way believing that anything has been developed as yet of commercial value." This assessment proved essentially correct.

Attached to this and other Du Pont reports on artificial leather surviving from those years are numerous samples, stapled to the original correspondence, or else rudely clipped to them with brass pins. A piece of split leather, the flattened knots of its fine, stringy fibers coated dull black. Swatches of fabric, candidates for covering with this or that formulation. New materials, experimentally treated. Some samples by now are crumbling or congealed, barely flexible. Some are oily and chemical, their ingredients leached into the papers to which they're attached, discolored and stained, the pale blue paper stock turned the color of tea.

Together, they are a testament to trying.

No one in coming years would look back to the work at Du Pont during this period and see any great breakthrough, like the Storey and Wilson patent of 1884, say, or polyvinyl chloride later. The record is more of frustration and failure than crowning success. For a while there was talk of using Fabrikoid for shoe uppers; they tried some out on local mail carriers, even sold a few hundred pairs. But the material didn't breathe, and the coatings were hard and brittle, cracking in use. Besides, the shoe manufacturers, burned by other leather substitutes, weren't receptive. And Fabrikoid never got much better. So for years Newburgh continued churning out pretty much the same old thing.

Later, Du Pont promotional materials stopped describing Fabrikoid as artificial leather at all, pitching it rather as an innovative coated fabric; it became "a pyroxilin coated textile," its manufacture "a triumph of chemical genius," its beauty harmonious with the steel and chrome modernity of the day. But that was later. At least up to the early 1920s, Fabrikoid's chief competitor, the standard to which it held itself, was leather.

One Du Pont ad imagined a meeting between a white-haired senior executive and his junior colleague. "Tom," the senior man orders, "let's bind it in Fabrikoid." Their new sales brochure had to "carry our message straight to the private office and the mahogany desk." Brochures bound in ordinary cloth wouldn't do, because they "don't look important enough." Leather itself was too expensive. But Fabrikoid would "get a leather welcome every time—you can't tell it from leather."

Some Dupont ads didn't mention leather, even as they hinted at Fabrikoid's superiority to it. One showed a men's club scene, two well-off friends in a circle of light and warmth, slumped in sumptuous armchairs.

COMPANIONS OF COMFORT

> The concord of true friendship, the fragrance of good tobacco, the enveloping luxury of Craftsman Fabrikoid—that's the comfort combination absolute!
>
> At home, at the club, in the hotel, wherever the deep chair and soft settee invite to restful ease, there Craftsman Fabrikoid is supreme. Rich in its coloring, caressing in its pliancy, thoroughly sanitary and serviceable, it is the aristocrat of upholstery fabrics.

No reference to leather.

Except that down at the bottom, next to the Fabrikoid logo, appeared a curious design—a hide laid flat, with truncated limbs, hints of tail and snout—the instantly recognizable symbol, then as now, for leather. Superimposed was a cow in profile and the legend, "How many hides has a cow?"

This referred to a Du Pont ad campaign, going back to about 1914, that emphasized how leather made into upholstery from a single cowhide could be made to cover triple the area of the hide from which it came. The secret was a machine that split hide into two or more layers; one version, developed before the Civil War, used a continuous loop of whirling band saw. Feed in a hide, which might be three-eighths of an inch thick, and *tztztzipp*, out came two hides—a top grain layer and another, legitimately "leather," but spongy, weak, and inferior in every way. Repeat the process and get a third, maybe even a fourth layer. As late as about 1905, they were still practically throwing away these lowly splits. But then tanners began coating and embossing them to resemble top-grain leather, sometimes with the grain of seal, crocodile, or any exotic species you liked.

And the coating they used? It was typically the same pyroxilin used by Du Pont to make Fabrikoid. By the time Du Pont bought Fabrikoid, split leather was common in American upholstery—and Du Pont was eager to point that out. "How many hides has a cow?" asked its ads; why, it was the *tanners,* they as much as said, who made inferior imitations! One Du Pont competitor pronounced it "the cleverest ad ever produced for artificial leather." And now in its "Companions" ad, by simply posing that sly old question, Du Pont was conjuring up those low-grade splits to which Fabrikoid was, presumably, superior.

In a 1919 article in its house magazine the company all but patted itself on the back for this promotional coup. "Through aggressive sales work and liberal advertising, the Du Pont Company has elevated leather substitutes, as a whole, from the weak position of an imitation of something better, to one of superiority over three-fourths of the leather made." Three-fourths was its concession to accuracy: Fabrikoid was not superior to the best leather, it was saying, only to splits.

But most Du Pont ads ignored such niceties. When, in 1914, the former Queen of Italy, Margherita of Savoy, had her limousine upholstered in Fabrikoid, Du Pont rushed ads into *Collier's Weekly* and the *Saturday Evening Post.* Here was evidence that Fabrikoid appealed "not because cheaper than hide leather, but because superior. Looks and feels like leather." Just leather, with no qualifiers. Sometimes, Du Pont called Fabrikoid a "manufactured leather," as opposed to "hide leather." Or else it was "an improvement on leather," in one case billed as offering "all the sumptuous richness and beauty of hide leathers." This was just the sort of talk that got Irénée du Pont nervous.

Fabrikoid was sometimes used for cushion covers, stage props, or other products not routinely made of leather. But more often it was leather's oldest, most traditional markets that Du Pont targeted—for example, bookbinding. "Gives the sumptuous appearance of fine leather binding at little more than the cost of cloth," Du Pont boasted in 1913. Law books, encyclopedias, *The Harvard Classics*—all had been bound with Fabrikoid. And when would-be customers did balk, it was typically because Fabrikoid failed to emulate leather in key respects. When an Experimental Station report surveyed the company's efforts to overcome bookbinder resistance, it cited Fabrikoid's oiliness to the touch and embossing problems—both of which made it look bad next to leather.

Simulate leather; that was about all Fabrikoid had to do. When, around 1921, the publishers of *Collier's New Encyclopedia* decided to bind its new edition in it, they adopted what Du Pont heralded in an ad as "a special process of embossing and hand coloring the covers in reproduction of a noteworthy specimen of exquisitely tooled Spanish leather." And yes, of course, as always, those covers were "practically indistinguishable from the work of the old master bookbinders."

So as Du Pont made and sold Fabrikoid in the dozen years after it entered the business, the natural material its technicians sought to mimic stood always before them. It was there in the company's sales pitches and clipped to its reports, at its plant in Newburgh, in the worries and preoccupations of its executives in Wilmington, under the eyepieces of its microscopes at the Experimental Station. Chemist Charles Arnold, who presided over much of the artificial leather work at the Station, began one 1920 report by telling how they'd gotten hold of a sample of leather upholstery from a Cadillac, "as a guide and standard of attainment." Years later he wrote a brief memoir about those days. In it he took care to credit his colleagues, among whom was one George Priest—"a valuable man," wrote Arnold, "because he knew much about real leather, which we were constantly trying to simulate."

Leather was the standard by which to measure their achievement, the natural material whose look, touch, and strength were the object of their quest, their Holy Grail.

3

Leather Alive

At noon on a late summer's day in 1834, 19-year-old Richard Henry Dana left behind the frock coat and kid gloves of a Harvard undergraduate, arrived at the Boston docks with his sea chest packed for a two-year voyage, and shipped out as an ordinary sailor aboard the small brig *Pilgrim*, determined to effect "an entire change of life." In his epic memoir, *Two Years Before the Mast,* he would recount his youthful adventure, much of it spent gathering cattle hides along the coast of California, when it was still Mexican. It was more than a year before the hides, bought from Indians with the flesh and fat still clinging to them, would reach Boston, 15,000 or more of them to a shipload. Young Dana's job, along with that of his shipmates, was to get them there unspoiled, ready to be made into leather.

Of the little coastal town of Santa Barbara, he recalled "the surf roaring and rolling in upon the beach, the white mission, the dark town, and the high, treeless mountains." There they collected hides in "the California style." Dried hides were brought down to the beach on mules or in carts. Then he and the others, each outfitted with thick caps, would swing one of the stiff, heavy, awkward things onto their heads, walk it barefoot into the surf, and throw it onto a small boat, which would ferry piles of them out to the *Pilgrim* at anchor in the bay.

Later, they bought hides from Indians gathered on the brow of a steep hill high over the sea near San Juan Capistrano. From that prominence—Dana guessed it was 400 feet high—they'd throw the hides onto the rocks along the shore.

> Down this height we pitched the hides, throwing them as far out into the air as we could; and as they were all large, stiff, and doubled, like the cover of a book, the wind took them, and they swayed and eddied about, plunging and rising in the air, like a kite when it has broken its string.

At the base of the hill, other men loaded them into boats.

San Diego was where hides gathered up and down the coast were brought to be stored before being loaded onto ships for Boston or other manufacturing centers. The trip around Cape Horn was long, and the hides had to be properly cured. Dana and his fellows would carry them to the water's edge at low tide, let the rising tide wash over them, leaving them there, to be cleaned and softened by the action of the seawater, for the next two days. Then they were rolled up, loaded onto wheelbarrows, and dumped into vats filled with brine, the natural salinity of the seawater fortified with great loads of pure salt. There they remained for another 48 hours after which they were stretched and staked, skin side up, still wet and soft. At this point Dana and company would fall on them with knives, kneeling on them, cutting out any lingering meat and fat—which, they knew, would "corrupt and infect the whole if stowed away in a vessel for many months." It was brutal work. It killed their backs; at first Dana could do only about eight hides a day. All afternoon, as the hides baked in the unfailing San Diego sun, grease would migrate to the surface; this, too, they'd scrape off. Then they'd unstake them, fold them over, hair side out, and leave them to dry.

There was more to it than that, not the least of it cramming the hides into the ships; "steeving," it was called. Ultimately, Dana would write, they'd be "carried to Boston, tanned, made into shoes" and other leather goods, some of them undoubtedly brought back to California "and worn out in pursuit of other bullocks, or in the curing of other hides."

Years later, in 1859, by then a well-known lawyer and author, Dana visited San Francisco, where he spied a pile of hides lying on the wharf. "I stood lost in reflection. What were these hides—what were they not—to us, to me, a boy, twenty-four years ago? They were our constant labor, our chief object, our almost habitual thought. . . ."

Today, meatpackers often tan hides, thus making them into leather, just hours after they've been pulled from freshly slaughtered animals. But in Dana's day, as for most of humanity's leathermaking past, hides were a separate commodity, with their own markets, transported by the carload or shipload around the world—raw material, not yet leather.

A hide is a hide and leather is leather. That, at any rate, is one truth.

Another, though, is that every piece of leather begins its history as a hide, hair and markings intact, bearing signs, veritable stigmata, of an individual animal's life. A Kenyan leather goods maker entering the world market with a new line of upscale handbags told a trade journal reporter how the camel hides they used hinted at the heavy scrub and thorn in the east Africa camel's habitat: "It looks like a camel that has had to deal with nature and lived in the real world." Whether an animal grazes in a meadow or is confined to a barn, gets sick, is fed too meagerly or too well, ingests chemicals, is branded or infested with ticks, is slaughtered in summer or winter—all these show up in the leather ultimately made from its skin.

Animals raised in warmer climates have shorter hair, their leathers showing up as smoother, with finer grain patterns. Hot days and cold nights may leave the leather with a "mosaic effect," the result of tiny underlying blood vessels that have become dilated. Poor diet yields thin skins lacking in elasticity. Leather's quality depends even on the part of the animal's body from which it comes. Sample a goat skin along the backbone and it might be 17 or 18 millimeters thick, around the belly 9 or 10. Moreover, the abdomen of an animal, like that of a human, stretches and sags, the corresponding leather stretchier than that taken from the back.

"Natural markings" is the way leather goods makers like to explain away wrinkles, scars, and other fleshly imperfections. In fact, leather with a clear enough complexion to show off in its natural state is rare and expensive. Most of what gets made into consumer goods needs to be buffed, embossed, pigmented, and otherwise treated to disguise blemishes. In these more frequent instances, in other words, the life of the animal registers *too* insistently in its hide to be blithely written off as charming variation or proof of natural origins. Each such defect makes the leather less valuable because of the need to cut around it, cover it up, or consign the hide as a whole to a cheaper product.

Jean Tancous *studies* defects. Back in the 1940s she went to work at the University of Cincinnati with Fred O'Flaherty, the leather chemist we met earlier who worried about the threat to leather posed by synthetics. She earned bachelor's and master's degrees in chemistry, has worked with leather and hides for the half-century since, gathering top awards from the American Leather Chemists Association and other groups along the way, and may know more about hide defects than anyone in the world. These days she works out of her own basement laboratory in suburban Cincinnati. Microscopes, ceramic dishes, plastic tubing, and small ovens share the space with throw rugs, a rocking chair, and hides offhandedly draped over stools. Shelves at the far end of the room are stocked with chemicals and books. Tanners and meatpackers send her hide samples; she reports back what she finds the matter with them. She'll slice out a little slip of hide, mount it in a steel contraption called a microtome, give it a blast from a cylinder of liquid carbon dioxide that instantly freezes it solid, and, with the microtome's sharp, precisely guided steel blade, cut a translucent, inexpressibly thin slice from the sample. Then she'll mount it under the lens of one of her several microscopes and look hard at it. Tancous is in her 70s now, her gray hair piled high atop her head with bobby pins. But when she's slicing hide samples or peering through her microscope, she displays the fresh, easy enthusiasm of a high school student working on a science project. Over the years she's seen it all, her knowledge collected in the second edition of *Skin, Hide and Leather Defects*, a 363-page compendium of animal hides gone wrong.

For example, disease—alopecia, eczema, ringworm, sycosis, warts: Each shows up in the animal's skin and affects the resulting leather. Ringworm lesions result in circular spots the size of a silver dollar; finish the leather all you like and the smooth, shiny spots characteristic of the disease will still be there. Likewise warts, which can cover half the skin of a cow with grotesque masses—hard, almost hornlike, invariably leaving a noticeable lesion that weakens the leather. Then there's lice, which bite, suck blood, or both. And grubs—maggots that make holes in the animal's skin so they can breathe. And genetic diseases, like "vertical fiber defect," seen in some Hereford cows, that yields leather so flimsy you can pass a pencil through it.

If a cow is scratched by cactus, thorns, or barbed wire, the resulting leather shows it. If it suffered as a hot branding iron scorched its skin, or if it was chemically branded with acid, or freeze branded with liquid nitrogen, the leather reveals that, too. Likewise if it was burned by hot tar or caustic paint on its way to the slaughterhouse, bumped or trampled in the feed lot, whipped or caned or clubbed or shocked by an impatient stockman. Even its final hours, just before slaughter, can leave evidence in the leather: Exhausted by being trucked from farm to slaughterhouse without sufficient rest and water, the skin overheats, the hide is incompletely bled and insufficiently cured, the resulting leather left subtly inferior.

Seen this way, then, the bond between leather and the once-living animal is indissoluble, and leather can seem a material just this side of living skin, its roots deep in nature and only incidentally the result of human effort—indeed, with a firm hold on that reliably fashionable label "natural." Listen to some leather goods advertising and it can seem as if the animal *is* its hide and *is* the leather made from it. "Some time ago on a cool, crisp October day, the idea of making shoes from American Bison leather was conceived by Harrison Trask during a day of fly fishing near his home in Montana." According to this brochure from a Montana-based shoe manufacturer, Trask noticed a herd of buffalo, meaning bison, grazing along the river's bank and wondered why he'd never seen shoes made from their hides. Pretty soon, his company was making shoes from bison, elk, and longhorn cattle; "American's original leathers," the company calls them in a trademarked slogan. A horned bison is part of the company logo. An elk with magnificent antler plumage appears in its brochure, to which is attached a small, chocolate-brown sample; "words and pictures alone cannot describe the soft, buttery 'hand' of Elk Leather," its qualities seeming to flow, naturally and inevitably, from the animal furnishing it.

Less exotic species, even ordinary cattle, are also sometimes hauled onto center stage on behalf of virtues they presumably bestow on leather made from their hides. A booklet issued in several editions over the years by the New England Tanners Club, *Leather Facts: A Picturesque Account of One of Nature's Miracles*, tells of "the meticulous care that ranchers give to their herds. It takes nature considerable time—the lifetime of an animal—to slowly and carefully weave together all the meritorious properties that ultimately yield a tanner's raw material." Great cattle make great leather.

At other times, however, leathermakers seem inclined to forget that leather comes from dead animals and are just as happy to let the link between the two wither or disappear. They'll tell you, for example, that animals are virtually never killed for their hides, that leather is an offshoot of another industry altogether, cattle raising for meat. In fact, they are invariably quick to tell you this; it's a persistent feature of the leather industry's perception of itself and appears almost without fail in publications representing its public face. *That other industry* slaughters animals, tanners as much as say; they, the tanners, simply retrieve the hides, otherwise destined to molder and rot, and put them to good use. "By purchasing this," reads the tag on a cosmetics bag my wife bought, "you are recycling. Tanning in the 21st century is considered a byproduct of the meat industry." *A byproduct.* All this is broadly true, but the 60 dollars a hide fetched in the summer of 2003 can spell the difference between profit and loss; sometimes you'll hear a hide called "the fifth quarter," suggesting income beyond that realized from the carcass's traditional four quarters.

Whatever the prices they fetch, the hides of the 35 million cattle slaughtered yearly in America alone have to be disposed of somehow, and they are; almost all become leather. As Mike Redwood, a British leather industry consultant, has noted, almost no animal hides, anywhere, at any time, wind up in dumps to putrefy; rather, a parallel, more or less smoothly functioning system exists, everywhere, at all times, for gathering hides and turning them into leather. Richard Henry Dana was part of it, on the coast of California, in 1834.

Leather is tanned hide; Dana's hides were merely cured.

Until tanned, a hide—the word typically refers to large animals like cows and horses, "skin" to smaller animals like calves, sheep, and pigs—risks spoilage by the same implacable biological forces that soon reduce an animal left dead in the wild to bones; bacteria produce enzymes that digest the hide's proteins, a process of putrefaction normally under way within hours of death. Salt stops it, drawing out moisture, making the environment inhospitable to bacteria. Salting a hide, or soaking it in brine, "cures" it; it's left in a state of suspended animation, temporarily preserved for its weeks or months of storage and transportation to a tannery.

But cured hide is not leather.

Rawhide is not leather, and neither is parchment.

All of these are hide that, left untreated and allowed to get wet, will rot.

Leather can sometimes serve as a breeding ground for mold, but it doesn't rot. "Imputrescible" is the word you see in dictionary definitions of leather, meaning not liable to decomposition, incorruptible. Leather doesn't need to be kept salty. It doesn't need to be kept dry. It doesn't need to be chilled. It can endure all sorts of biological, meteorological, physical, and chemical abuse; made into shoes, saddles, and belts, it routinely does. As hide it might decay within days. But as leather it can last, without special care, for centuries. At the Museum of Fine Arts in Boston you'll find a handsome coat made in Egypt from white antelope leather in about 3000 B.C., pretty much good as new. In 1973 a Danish brigantine, the *Frau Metta Catharina*, was recovered where it had sunk almost 200 years before, near Drakes Island in Cornwall, England. Recovered with it was a shipment of reindeer leather, tanned in the Russian style, with willow bark, bound for Genoa. Most of it was none the worse for wear. Some was made into a pair of shoes for Prince Charles and today, on the Internet, you can buy belts and notebook covers made from it, complete with a certificate of authenticity.

Tanning is what makes leather *leather*, gives it permanence. And in halting those biological processes that would otherwise have caused the hide to rot, it severs it from its roots in nature; in this narrow sense, certainly, it is hard to credit leather as "natural." How natural can it be, how beyond human artifice, when the same hide can equally well become a soft, drapy vest or a hard, crusty instrument case that looks as if it could stop a bullet? As we'll see, complex chemical and mechanical processes, developed, refined, and perfected by generations of tanners, transform hide into leather, and so erect a kind of conceptual wall between them; they are so alike, yet so fundamentally different. Yes, leather has qualities over which its enthusiasts gush. But as it comes from the tannery—which is, of course, a factory—leather is a product stable and consistent enough to be listed in industrial catalogues broken down by type, thickness, color, and shade. Its density, tensile strength, and water absorption can be found listed on specification sheets right beside vinyl sheeting or nylon filament. It is, in short, as much thanks to the miracle of tanning as to the miracle of life that leather is so stable, strong, beautiful, and loved.

"It's still unique after all these years of man-made fibers, and I think it will continue to be unique for hundreds of years to come." Nick Cory, a tall, smooth-talking Brit, a Ph.D. chemist, is referring to leather. Cory is director of the Leather Research Laboratory outside Cincinnati and host of a twice-yearly leather orientation course offered by Leather Industries of America, a trade association. The two-day course takes participants from the biomechanics of skin and leather chemistry to test instruments and Federal Trade Commission labeling requirements. On hand for it are new tannery employees, a buyer for a company that makes saddles for Harley-Davidson, two from a Mexican supplier of leather steering wheel covers, and half a dozen others, including me.

It's the first day now and Cory is lecturing about skin, its hair follicles, fibroblasts, collagen networks. To help with his anatomy lesson he flashes on the screen an image of skin in much-magnified cross-section. Hairs erupt from the surface. Their roots reach down into follicles, are fed oil by sebaceous glands. Little muscles, the erector pili, make hairs stand on end. A network of fine veins and capillaries feed blood to the area. The skin of Cory's cross-section, as it happens, is not that of a cow, pig, or deer—it is human. And his image choice is neither lazy, ghoulish, nor misleading; all skin is about the same.

Now, *certainly* the skins of animals differ; the differences show up in the leathers made from them, in their tear strength, their breathability, their fitness for binding books or lining shoes. Before me lies a photocopy of a United Nations report on novelty leathers, like alligator and ostrich, images of several appearing on the cover. The photocopy is poor, scarcely apt to reveal subtleties of form and texture. But even so, they're each as different from one another as a Rembrandt is from a Rauschenberg. Likewise more everyday leathers, though you might want a magnifying glass to spot the subtler differences in hair cell pattern between fine-grain calfskin, say, and ordinary cowhide. Pig skin has large hair follicles, arranged in groups of three. The skin from the rump of a horse makes for a tough leather, cordovan. Elephant hide, occasionally used for certain specialty leathers, can be an inch and a half thick. Goat skin makes soft leathers, with beautiful drape, for garments. So plainly, it's wrong to collapse the skins used for leather, in all their variety, into just plain "skin."

And yet not *very* wrong.

You can make leather from just about any animal. Kangaroo, elk, shark, cow. A Massachusetts firm started up to sell leather goods from salmon, cod, and wolffish. *Do they smell?* its proprietor, Jim Bates, is often asked. "Well, does cowhide smell like cow?" he asks right back. Frog's legs, turkey legs, beaver tail—all, he notes, can be made into leather. At the leather orientation class, another instructor, Randy Rowles, told how he guided his young son through a science project in which they tanned a mouse skin in a tea cup. "We got to it before the cat did," he says.

At the Smithsonian Institution in Washington, D.C., they just might show you their boot made of human skin. They're touchy about it, of course, and don't much like to talk about it, but there it is, in their collection. It seems that back in 1876 the upcoming Philadelphia Exposition, celebrating 100 years of American independence, was deemed *the* place to industrially and commercially show off. Everyone wanted in, including a New York City shoemaker, H. and A. Mahrenholz. Their firm, they wrote the Smithsonian, would be happy to make a pair of display boots from alligator leather. Or from anaconda skin. Or, if the Smithsonian liked, from human skin. Their boots, they promised, would prove "very handsome strong and durable, so weather or time of years would not effect them." Sure enough, they were soon shipping all three, "trusting that you will give them a prominent Place in the Centennial Building," and adding a postscript: "Find enclosed 2 pieces of the Human Skin same as the Boots are made of."

The Smithsonian wrote back that it wanted to know a little more about the humans whose skin went into the boots. The Mahrenholz reply came back a little fractured—they were shoemakers, after all, not grammarians—but clear enough:

> Its from two male subjects old men, off the Stomach or rump to the back bone. If its a question wheather you shall dare to exhibit them as such after the pains we took in making them for that purpose only, merely, to show what could be made out of human skin as well as any other skin. They are not made to wear, which of course you understand, but only for the novelty of the 19th Century, something that is never been exhibited before. Women might shrink from seeing them, as well as some men yet do not! Our most refined fashinable women wear the hair of dead Persons on their head.

The boots are in fine shape 125 years later, light brown to the eye, soft to the touch. But otherwise there's little to say about them: Run your hands over them. You won't cringe. They're just like any others.

So leather, roughly speaking, is leather.

Leather comes from skin, but not the whole skin.

Your skin, or a cow's, or a deer's, has hair, or fur, the hair roots reaching deep below the skin, into oily follicles. At the very surface, the outermost layer, so papery thin it's apt to degenerate into a bare, scarcely discernible line in even a highly magnified cross-section, is the epidermis, which grows cells and sheds them and is never more than a few cells thick. It and the hair passing through it are discarded in making leather. So is the bottommost layer, attached to underlying fat and flesh.

What's left, the middle and usually thickest layer, is what becomes leather. It consists almost entirely of two sublayers: the fine fibered grain layer, bearing the skin's characteristic pattern of hair follicles, and below it the corium, a dense matted tangle of slightly coarser fibers. This, together, is the material that, from the grain side, you might admire for its buttery touch or, from the flesh side, you might inspect to confirm it's the real thing. Much of the tanner's work goes into getting everything else out of the way so he can reach, chemically and mechanically, this middle layer.

If you're young, with good eyes, you might be able to discern leather's fibrous tangle with the naked eye. For the rest of us an ordinary magnifying glass helps. Examine a thickness of leather, one cut cleanly by a sharp knife, and you can see them, the fibers clipped short, like a man's crew cut, but some seeming to culminate in short stunted tufts. Our leather sample, remember, has been *cut*, its individual fibers cut, too. But if you dig down, deep into the matte of fibers, you might be able to follow one that's escaped cutting, one passing in, out, and through the tangle; that's what Randy Rowles did once—teased apart from its neighbors a single fiber of leather, followed it for its entire length, and wound up with a piece a foot long. Look more closely yet at any such fiber and it's composed of still finer fibrils, finer by far than human hair. Once I made a scratching post for our cat from a wooden post wrapped snugly by a piece of thick leather, suede side out. By the next day Herky's eager claws had transformed its peach fuzz of suede into a loose eruption of fine fibers, some of them three-quarters of an inch long; together they were like a man's hairy chest; each, individually, like a dandelion tendril trembling in the breeze.

These fibrils are made mostly of collagen; think of leather as simply collagen and you won't be far wrong.

What cellulose, in its ubiquity, is to the vegetable kingdom, collagen is to the animal world—all over, everywhere, the single most abundant protein in animals. It forms tendons, ligaments, and cartilage. And it makes up by far the largest constituent of skin, whether of fish, human, or cow. Collagen fiber itself is strong—pound for pound stronger than some kinds of steel. Twisted and matted together as it is in leather, intertwined in every direction, and you wind up with a material, beyond merely strong, whose look and feel people have tried to imitate for hundreds of years.

Boil hide, bone, and connective tissue and you get gelatin, the prime constituent of Jell-o, jams, and certain glues. When, in folktales, Ol' Gray is hauled off to the rendering plant at the end of his life, it is gelatin he becomes; collagen, a word apparently coined about 1865, means "yielding glue." H. R. Procter, the British doyen of leather, wrote in 1903 that "so far as our present knowledge goes we may regard hide-fibre as merely an organised and perhaps dehydrated gelatin." That view, wrote another British leather scientist most of a century later, "does not depart too greatly from the picture today, except that we should now regard the gelatin structure as derived from that of collagen rather than vice versa." Collagen, in other words, comes first.

In 2004 the American Leather Chemists Association held its annual meeting in St. Louis, where it had held its first meeting a century before. There, in the city's Chase Park Plaza Hotel, Dr. Jaume Cot, of Barcelona, president of the International Union of Leather Technologists and Chemists, gave the association's annual honorary lecture. Its title was "An Imaginary Journey to the Collagen Molecule," nominally devoted to prospects for better tannery waste treatment but going much further afield. At one point he played a harp recording of a musical composition derived from the amino acid sequence of the collagen molecule. Cot knew his audience. Collagen is the leather chemist's prime object of scientific interest. It's most of what any hide is. It's what the tanner spends his waking hours trying to transform.

For centuries the technology of leathermaking changed with glacial slowness. The cured hide came in, you got the hair off, you tanned it, you dyed it, you rubbed greases and oils into it, you shipped it out; that's essentially what tanners do today and pretty much what they've always done. For the past two centuries or so, chemists have helped tanners better understand what they were doing. But leathermaking long preceded scientific understanding. For the most part, it was improved, when it was improved at all, through everyday practice; science was but little part of it. *So* little, to cite one example, that at the London Crystal Palace Exhibition of 1851, all aflutter over the McCormick reaper and the Colt revolver, exhibition judges dismissed recent patents and processes devoted to tanning, saying, "There has been no decided improvement, no marked progress, to show that better results have been obtained than by the old methods." Leather was tanned as it had been for centuries. And tanning's technological instrument, for most hides and most leathers, was nothing more exotic than bark, leaves, and berries.

The story goes that one day in the Gran Chaco, the great lowland plain east of the Andes in Argentina and Bolivia, a thunderstorm broke out and lightning struck a tree, killing a cow that had taken shelter beneath it. Some time later, the dead cow was found in a pool of brown water, its hide preserved as leather. The tree was a quebracho, which makes a hard, heavy wood; *quebar hacha* means "the hatchet breaks." The thick corky bark from this tree turned out to be one of the best vegetable tanning agents, with among the highest concentration of natural tannins.

But quebracho has been used for only about 100 years. Before that, and since, many other vegetable materials were used, typically whatever was native to the area. Pine and mimosa were among the trees furnishing tannins. So were spruce and gambier. In the American colonies it was hemlock in the north, oak in the south. A chapter devoted to vegetable-tanning materials in Procter's century-old classic, *Principles of Leather Manufacture*, could be mistaken for a botany textbook, studded as it is with Latin names like *Castanea vesca* and *Rumex hymenosepalum*, photographs of hemlock fir, and line drawings of Sicilian sumach; any of them could make vulnerable hide into robust leather. But large quantities were needed; part of most tanneries was a bark mill, to crush or grind the raw material into useful form. A Civil War-era account tells of one with "a circular trough fifteen feet in diameter, in which the bark was crushed by alternate wooden and stone wheels, turned by two old blind horses, at the rate of half a cord a day." Drawings, patents, and other records suggest that this description could have applied, virtually intact, for at least the previous two centuries.

In times past, making leather could take a year, or even two. You placed hides into a pit, grain to grain, the flesh side facing out to a layer of bark, leaves, or nuts, layer upon layer, tanning agents gradually diffusing into the hides. Later, tanning infusions or liquors of various strengths were used, each an extract drawn from the raw materials. In time tanners learned that stronger tanning solutions, or "oozes," left a hard, impenetrable outer surface, inhibiting tannage of the interior. So you needed to go slow, changing the tanning liquors in a set pattern that only gradually raised the acidity and strength of the solution, or else moving hides from pit to pit of progressively greater strength.

You had to be patient, and tanners eager to move out their goods and get paid weren't always. The preamble to the acts of Edward VI, dating from before 1553, records that whereas hides were normally to remain in pits or vats for a year or more, some tanners had "invented diverse and sundry deceitful and crafty means" to tan leather in a month. One was to put "seething hot liquor with their oozes into their tan vats," something apparently done surreptitiously at night.

Tanning is the essence of leathermaking, so much so that this one iconic word is sometimes indiscriminately applied to the whole process: *Tanners, working in a tannery or tanyard, make leather.* But not all of what they do is tan; leathermaking comprises much else, and it was often in these other steps, some preceding the tanning itself and some following it, that its artistry resided.

Before a cured hide could be tanned, the first step was to thoroughly wash it free of blood, dung, and dirt and, in so doing, restore natural moisture lost to the salting. This usually meant immersing the hide in the local stream. At some point, the flesh side of the hide would be scraped clean of any remaining muscle or fat.

Getting the hair off the grain side of the hide typically meant dumping it into vats of lime (which, mixed with water, became calcium hydroxide, a strong alkali). That loosened and dissolved the hair sheathes, so that the hair, still intact, could be scraped off with a blunt knife; the British termed the resulting mess of lime and hair "scud." Later, tanners learned they could avoid much scraping by adding sodium sulfide to the process, either as a kind of soupy solution poured over the hide, or in a vat into which the hide was suspended. Either way, this caustic, extremely alkaline solution attacked the hair itself, resulting after a few hours in a paste, sans hair, that could simply be wiped off. It smelled terrible, like rotten eggs; the chemical reaction produced hydrogen sulfide.

At this point the hide was free of hair but full of lime, almost unrecognizably swollen, perhaps twice as thick and heavy as before, hair follicles prominent. Squeeze it between your fingers and it would take up their imprint. The next step, then, was deliming, which flooded the hide with salts like ammonium sulfate, reducing its alkalinity and the swelling that accompanied it.

Accounts of traditional leathermaking sometimes take perverse delight in the foul, disgusting materials on which it's relied over the centuries. These include urine, feces, stale beer, and rotting scraps of hide. Many of these colorful, not to say pungent, ingredients find use in "bating," an ancient practice that has lent something of its aura to the leathermaking process as a whole. Bating, and its near cousin, "puering," meant using warm excretory materials—14 quarts of dog dung to every four dozen skins, by one account; diluted to the consistency of honey, by another—to leave the hide better ready for tanning. It made the hide softer, too, in part by destroying residual hide constituents, such as hair roots, pigments, and gluelike protein substances. "A peculiarly foul and repulsive process," someone once called it. But it worked. As they emerge from bating, according to one early 20th-century description, the skins "have lost their rubbery condition and now appear as soft and flaccid as a silken gown."

The result of these preparatory steps was a hide that, in a sense, was *essence-of-hide*—a purified, almost sanitized collagen network, largely hairless, fatless, and without extraneous proteins. Everything was gone, in short, except what the tanner had coveted from the beginning—those fibrous collagen tangles, unencumbered by anything else to get in the way. Ready, finally, for tanning.

Yet "finally" gives the wrong idea, because the steps that followed tanning were as important as those preceding it. After its year in the pits, the tanned hide was now leather, yes—but leather no one, then or now, would want in a pair of boots or piece of luggage. A hide had been extracted from the world of growth and decay it had inhabited not long before. But now what in the living animal had been soft, pliable, and resilient was left as "crust"; that's the traditional term, and indeed the leather *was* now hard, crusty, and unresponsive to the touch. In a sense, tanning had sucked the life from it. So the idea now was to give it back something of that living softness and tactile warmth it had once enjoyed.

This, traditionally, was the work of the currier; in French the word was *corroyeur.* Both words had their roots in *corium,* Latin for leather, the word today given to the middle layer of hide made into leather. Curriers dressed, dyed, and otherwise finished the tanned hide. A shoemaker in medieval England, for example, might buy tanned hides from the tannery, only then to hand them over to a currier with careful instructions about how he wished them finished, or "dressed," typically for greater softness and pliability. Currier and tanner were almost never the same person; in England they ordinarily belonged to separate guilds.

The currier used something like the tanner's familiar inclined beam to scrape away any extra flesh from the leather. But that was about the only tool they had in common. The currier used special thumb-shaped hammers and screens to beat, scrape, stretch, and pummel the oiled and greased leather into softness. He shaved it down to the right thickness. He impregnated it with fish oils, waxes, and dyes; all these processes have their counterparts in mechanized operations today. In Pope's translation of Homer's *Iliad,* the Greeks and Trojans fight for the body of Patroclus:

> As when a slaughter'd bull's yet reeking hide,
> Strain'd with full force, and tugged from side to side,
> The brawny curriers stretch; and labour o'er
> The extended surface, drunk with fat and gore.

England in 1553, Germany in 1747, America in 1860: Glimpses of leathermaking from those places, across 300 years, don't seem so different: "The tanning and dressing of leather," wrote John W. Waterer, an English student of leather, in 1946, was among the last occupations to leave the middle ages behind. "Until the end of the nineteenth century there was little, if any change in the methods used for centuries." Most characteristically, leathermaking was the beam, leathermaking was wet, and leathermaking made for a godawful stink.

Leathermaking was the beam, the long, inclined piece of wood over which workers toil in virtually every depiction of leathermaking from the middle ages until recently. Take, for example, Nuremberg printmaker Jost

Amman's illustration of a tanner, dating to 1568, in Hans Sachs's classic book of poems, *All the Trades on Earth.* Ox hides, complete with horns and tails, hang on a rafter above him. Behind him appear the half-timbered homes of a German town. In the foreground stands the tanner himself—aproned, thigh-high boots delivering him from what looks like filthy, festering tannery slop, with shirtsleeves rolled up, revealing thick, muscular forearms. He is scraping flesh and surplus tissue from a raw hide, using a two-handled fleshing knife. He reaches down over the hide, putting the full force of his arms and shoulders into the task, as it lies draped over—and here is the symbol, the icon of tannery work through the ages—an inclined beam.

"Working the beam" is the English caption in one edition of the Sachs-Amman project. Sometimes the beam was just a length of tree trunk, sometimes a board gently rounded along the top. You see it in Jost Amman's iconic drawing. You see it in Diderot; in French the beam is *chevalet,* or saw horse. You see it even in photos of tannery operations from the 1940s or 1950s, men leaning over their beams, arms outstretched, reaching into the flesh of the hide with their knives, an embodiment of male physicality. These days the beam itself has disappeared, but the whole range of leathermaking work involved in preparing hides for tanning is still called the work of "the beamhouse."

Leathermaking was wet. Leather in use is normally dry, but it gets that way only near the end of the leathermaking process. At least until the final finishing or currying operations, every tannery operation meant treating hides with one or another rinse, solution, "liquor," or "ooze," all of which kept them wet. Sometimes merely moist to the touch. Often sopping wet, soaked, saturated, the hides heavy, awkward things, that slapped noisily onto the tannery floor, the men tending them garbed in aprons, boots, or the other foul-weather gear of their era.

"I dry the skins out in the air" begins one translation of the Hans Sachs poem:

> Removing first each clinging hair
> Then in the Escher stream I dash them,
> And thoroughly from dirt I wash them.
> Cow-skin and calf in tan I keep,
> Long months in bark-soaked water steep.

Tanneries were invariably situated beside a flowing stream in which the hides were cleaned. In times past, they comprised vats, tubs, or long, stone-lined pits, the hides sitting in them for months on end. Nowadays, they're sloshed around in great wooden drums, then dumped out as great floppy masses of heavy, wet hide. Past generations had different names for, and different understandings of, just what happened to the hides at each stage, but always they were being washed, soaked, steeped, softened, or inspirited with one or another magic potion, acidic or basic, toxic or benign, but always wet.

And then, of course, *leathermaking stank.* Everything about it helped make tannery work what one scholar termed "odious, dirty, and no task for an aesthete." It started with a raw material of hides and skins taken from slaughtered animals that within hours had begun to smell, and in the frequent absence of reliable curing continued to smell, and soon smelled worse. The water used to clean hides comes out full of blood, fat, and dirt. Until the middle of the 19th century, at any rate, hides were typically brought from the butcher with hooves, horns, and other appendages attached; "as a result," it's been noted of early British practice, "a heap of rotting offal soon collected in most tan yards." Indeed, much historical knowledge of tanning owes to surviving records of edicts drawn up to control tannery nuisances.

Bad enough was the raw material, always threatening to rot and stink. But it got worse. One common means of dehairing a hide consisted simply of folding it up and letting it start to rot. You could speed up the process by sprinkling urine over it. And then in the beamhouse there was all that dog or pigeon excrement in its sheer animal putrescence. The protagonist of Patrick Suskind's novel *Perfume,* set in 18th-century France, goes to work in a tannery at age 8 and is assigned the most disgusting and dangerous tasks. "He scraped the meat from bestially stinking hides, watered them down, plucked them, limed, bated and fulled them, rubbed them down with pickling dung," and so on, the odious smells of the tanyard taking root forever in the twisted mind of Suskind's character, a serial murderer.

If we stop here, with this graphic depiction and these sharp pungent smells, leather's roots seem indisputably in nature, any chemical or mechanical steps involved seeming for the moment secondary or peripheral, leather itself virtually a synonym for skin. In a 2004 promotion in

Paris, "Natural Sensations," a consortium of Italian tanners honored its line of vegetable-tanned leathers with a kind of verse, picturing it as the work of Nature at its highest and best.

> Nature gives us hides, water, and through plants, tannin,
> Nature gives us mind and muscle.
> And so, vegetable-tanned leather is born.
> Perfect just as it is.

Indeed, the way leather was made all through the centuries—think of good, simple men with strong backs, working with hides cut one by one from once-living animals, toilers in barks, berries, and leaves, out in the open air, beside the river, the smells of life and death—was this not the very embodiment of "naturalness"?

I must tell you, however, of a number of complications, contradictions, and alternate readings to this charming tableau.

Take, for example, bating, which can seem to epitomize the very most "natural" in leather. Late in the Victorian era, English chemist Joseph Turney Wood began to rethink bating, long shrouded in mystery, as an industrial process like any other, susceptible to human understanding, rooted in science, reducible to chemical formulas. In 1886 he went to work in a tannery, learned bacteriology, and identified microorganisms in dog dung, whose chemical composition he studied. In an 1894 paper he concluded it was still "impossible to produce commercially an artificial bate" the equal of dog dung. But its mode of action was hoving into view: Bacteria secreted enzymes that broke down the hide proteins lingering from earlier stages in the leathermaking process. In papers appearing in 1898 and 1899, Wood suggested how a bacterial culture might do the work of excrement. His vision took material form in 1908, when German chemist Otto Röhm introduced a powdered, enzyme-rich concoction, "Oropon."

Here was *Biotech, circa 1908*: the product of a lab, complete with brand name, a clean new product remote from the biological refuse that had done the work for eons. Wood's 1912 book-length survey of bating ended with an apology that his subject, which was, after all, excrement, was "not an inspiring and lofty one," but expressed hope that one day "the use of such filthy materials may be entirely avoided." Today, leathermakers still speak of bating, still do it to leather, but don't any longer scour kennels, buying the needed materials from chemical supply houses instead.

Tanning itself, beginning as early as 1794, during the French Revolution, also began to seem less nature's work and more man's. Soon after chemist Antoine Lavoisier lost his head to the guillotine, his student, Armand Seguin, reported that you didn't need bark to tan leather. Properly mix the bark with acid and water, and the resulting solution would contain tannins—this was Seguin's word for whatever was responsible for the tanning, the active principle—that could be used instead and to better effect. No need to fold up hides and leave them sitting in pits with ground bark for months at a time. It was enough to immerse them for a few days in these liquors.

Seguin's innovation actually represented a high-water mark for him in what might otherwise have been a brilliant scientific career. Because it contributed so substantially to the manufacture of shoes, saddles, and other gear vital to a nation at war, which Revolutionary France was, Seguin became wealthy, and his wealth deflected him from science. But his one big discovery was enough to turn tanning away from its ancient roots in nature. For if you looked at what was in Seguin's vats, it was no longer bark or leaves but something *extracted* from them—at one remove now, a little more abstract. A few years later, in 1805, inspired in part by Seguin's work, English scientist Charles Hatchett went before the Royal Society of London to claim that by digesting coal, turpentine, and other carbon-containing materials, he could make substances "very analogous to tannin," going so far as to tan some skin with them. His paper's title: "On an Artificial Substance Which Possesses the Principal Properties of Tannin." He, Seguin, and others were, around the turn of the 19th century, making leather a little more the work of human artifice—which is to say a little more "artificial."

Seguin had shown that bark extracts might do as well as the bark itself. But in 1884, in Philadelphia, came proof, through a major technological and commercial innovation, that you didn't need berries or bark, didn't need to fell whole forests, didn't need the vegetable kingdom at all. Back in 1797 a Frenchman had identified the chemical element chromium, found in ores mixed with iron, that you dug up from the ground, cold and hard, then extracted from its ores with powerful chemicals. This metal would soon largely replace the oak and mimosa, chestnut and quebracho, that had been used to make leather for ages past.

Nonorganic methods of tanning were not entirely new, just not widespread. Alum tanning, or "tawing," went back to the earliest days of leathermaking, often being used for goat skins. It used potash alum, a sulfate of aluminum and potassium, typically in the form of fine white crystals; one recipe used egg yolk, flour, water, and salt along with the alum. The resulting leather was stretchy and vulnerable to water. But alum tanning did make for a unique snow-white leather and had its niche.

In Philadelphia, in the 1880s, it was much used to imitate prized Morocco leather, which was traditionally made from kidskin, tanned with sumac extract, and finished with a waxy coat. But when someone tried to use alum-tanned leather to sheathe steel corset stays, the stays rusted and the leather stained. One Philadelphia tanner turned for help to Augustus Schultz, a chemist in his 40s who'd immigrated from Germany while still a boy and worked for a nearby dye firm. Could he devise some method that avoided the problem?

Details about Schultz are sketchy, but the results of his work are embodied in two U.S. patents that were issued in 1884 and challenged in a series of infringement suits over the next 15 years, though not successfully. Chromium was in the air; its salts had indeed been tried earlier on hides. But nothing commercially had come of those efforts, in part due to the resistance of tanners to anything but the familiar look, color, and feel of traditionally tanned hides. One historian of this epochal episode in technological history, Joseph J. Stemmech, dubbed those years the "chrome tanning awareness period," during which chrome failed to supplant the old ways.

But now, with Schultz, it did supplant them.

Since its discovery, chromium compounds had been a mainstay of yellow pigments and dyes, and pigments and dyes were Schultz's area of expertise. The heart of the Schultz patent lay in the production, within the vats, of chromium sulfate, which as an industrial chemical today comes as a soluble green flake or powder. Indeed, the mark of a properly chrome-tanned hide is an instantly identifiable blue-green hew that reaches deep into it; today, "wet blue" is the universal name for leather that, chrome tanned and kept moist, is ready for subsequent finishing. At any moment today, thousands of piles and pallets of wet blue litter the tanneries and cargo holds of the world.

Tanners learned early on that chrome-tanned hides had to be kept wet; they could not be dried to a "crust," then rewet for subsequent finishing, as veg-tanned hides could. A chrome-tanned hide allowed to dry out is hard, unyielding, virtually impossible to do anything with. Drop it on the floor and, rather than flop over with that easy languor we impute to leather, it will smack down noisily, more like heavy cardboard or even wood than leather. Fold it back on itself and its surface is apt to tear, the grain layer delaminating from the underlying corium. Chrome tanners soon learned to immerse their wet blues in an emulsion of oil and water—neatsfoot oil or olive oil with Castille soap was one early recipe—letting the oils, fats, and their more modern equivalents seep into the wet leather. In effect, the natural lubricating effect of water was replaced with that of fats. "Fatliquoring" this process came to be called throughout the industry—the term is apparently unique to leathermaking—and it's a crucial part of leather manufacture to this day.

Fatliquoring is today part of the broader "retanning" stage in the making of chrome-tanned leather. Strictly speaking you don't tan the leather again, at least in the sense of staving off putrefaction; it doesn't need that. What it does need is improvement in its physical qualities—to make it softer, easier to dye, or in the case of overly blemished hides, easier to buff and emboss with a new surface grain. Once, old leather hands would say, "Leather is made in the beamhouse," meaning that the preparatory, pretanning steps was where the art lay. But now you'll hear that the artistry comes in retanning, fatliquoring, and finishing.

A chrome-tanned hide differs a little from one that's been vegetable tanned. Veg tanning yields leather that, depending on the vegetable extracts used, is typically buff colored, or brown; variation beyond that limited range is more difficult to achieve. Chrome tanning yields first its characteristic blue-green, which can be bleached in retanning and then dyed any color you choose—lilac, chartreuse, or Cardinal red as readily as the traditional beiges and browns. Veg tanning is better for hand tooling, tends to stretch less; you see it a lot in tack shops, where saddles, harnesses, and other equestrian gear are sold. Chrome-tanned leather is typically (though that word is misleading given the range of the tanner's art) spongier, softer, and more colorfast.

Mainly, though, chrome-tanned leather is cheap.

Chrome cut the time needed to tan from weeks to hours, expanded the scale of the industry, and made Augustus Schultz a fortune. At a time when leather workers made $12 a week, Schultz sold his patents in 1891 for $50,000, the equivalent of more than $1 million today, and retired to a Pennsylvania farm. The company buying the rights built a new tannery in Philadelphia, described in a pamphlet it issued as "the largest plant of its kind in the world." Eleven buildings spread over 14 acres, 1,600 employees, 40,000 skins a day.

Once it might have been easy to see leather as a self-evidently natural product—cows and sheep its agricultural raw material, dog dung essential to its making, the tanning agent itself drawn from neighboring forests, the bark or other familiar vegetable materials they yielded ground down to usable form right there at the tannery, the human imprint on it light, anything like "science" or "technology" irrelevant. But by the eve of World War I, certainly, in the wake of artificial bates, the Schultz process and its industrial cousins in chrome, that nostalgic, preindustrial view of leather was more difficult to hold.

So is leather best imagined as the skin of an animal that once ate, breathed, and bled? Or as an industrial commodity like paper, cotton, or brass, made in a factory, traded in the marketplace, fixed by technical specification? It's both, of course, the barest shift of perspective making it seem more one than the other. Consider a product labeled GENUINE COWHIDE or, on the other hand, GENUINE LEATHER. Each reflects our gaze in ever-so-slightly different ways, the first toward the natural world, the second toward human industry.

Today, leather chemistry is a well-established science. Most leather (about 85 percent in the United States) is chrome tanned, a technology little more than a century old that's about as "natural" as steel making or oil refining. The industry's technical journals are filled with formulas and flow diagrams, syntans, triple helices, glutaraldehyde linkages, and copolymers. Workers at consoles monitor computer-controlled operations.

Yet take your eyes off those whirling tannery drums for just a minute and the *life* in leather can get its hold on you again. Indeed, as hides pass through the tannery, it is the living skin that often serves as model, or standard, for what the final leather should be. "The skin, in its living condition on the healthy animal," wrote Joseph Turney Wood in 1912, "is the most supple and perfect of coverings, and in producing soft leathers it is the object of the tanner to retain this supple condition."

It is possible to carefully fold and stack hides so that they affect something like the orderly neatness of other industrial materials. But more often, around the tanneries of past times as well as those of today, at every stage from slaughter to finished leather, the hides are there in all their wet, floppy irregularity, bearing vestiges of feet, tails, and heads, the drums and conveyors laden with them overfull with the wrinkles and veins and messy, misshapen protuberances of life. Then it's harder to entirely forget that what tanneries produce, what workers slice and shave and move from pit to pit or drum to drum, what men and women make into wallets, belts, and shoes is, in the end, the skin of a once-living animal, endowed with something of life still in it.

We are deep in the American Midwest, at a "raceway," a large, oval circulating pond, maybe 60 or 70 feet long and a dozen wide, a churning industrial river of salty, frothing water and close-packed cowhides. It is hard not to think of a washing machine, the kind with a glass-fronted door, where you watch your clothes being pummeled in the soapy water, suds sloshing over them, a leg of blue jeans swinging into view, then a twisted clump of red flannel. But here it's not soapy water but dirty, brown, blood- and manure-filled brine, not your week's wash but hundreds of heavy hides.

They're fresh from the packer, where the animals were killed, suspended from chains, their blood drained, their heads and hooves cut off, their hides pulled or cut from the carcass. Then the hides were trucked here and dumped into the raceway, where now two great rotating wheels reminiscent of a Mississippi paddle wheeler ceaselessly propel them around the oval. Bright red stockyard tags, still attached by adhesive to some of them, occasionally surface against the backdrop of brown. As they curdle and slosh, now a hide's flesh side peeks through the froth, now the hair side with all its old familiar animal markings—those gently rounded patches of brown, black, and tan, icon of bovine serenity.

For 16 hours the hides circulate, making a river of hair, skin, flesh, excrement, water, salt, dirt, and blood. Then they're hung on a moving conveyor and taken to be fleshed, where they're fed one by one into a machine whose big, brutal cylindrical knives stretch the hide, grind off the fat, and roll it back out, where men with knives trim away irregular little pieces of useless hide; a large wall chart shows a hide with sections marked to remind them what they're to cut away and discard. Another man at the end of the line grades the hides on the basis of weight, condition, and defects, before they're piled up and stored.

Most of the place is more warehouse than factory. Everywhere are pallets and small trucks and containers stacked high with hides, some of them those of sows, sprinkled with salt, neatly folded up, like a shirt come back from the cleaner, an irregular and complex shape reduced to regular and predictable folds, only with little shreds, snippets, and lumps of flesh hanging out. The hides, maybe three-eighths of an inch thick, are wet and heavy, with enormous substance and bulk; you can't help but be impressed with their sheer physicality, and their weight, 100 pounds or more each; there's nothing here to remind you of the fineness and delicacy of thin finished leathers, nothing in common with the feel of a fine wallet or a sexy little black leather skirt. They are thick, greasy, slippery, and cold; lift up a hide from the top, let it go, and it flops back wetly, heavily, onto the rest of the pile, like a wet towel on a shower room floor.

When the market price is right, these hides, the skin of cattle, will go to Mexico or Taiwan, where they will begin to be made into leather.

4

Bizarre Effects

It's yucky stuff, I might as well tell you. This particular swatch of upholstery material is supposed to remind you of a palomino, with its golden coat, maybe one that's just clambered up from the river, thick hair clinging to its body, wet and matted. COMANCHE, it's called, and it's made of vinyl. Back in the 1950s, they used it to cover the chairs and bench seats of the Saddle and Sirloin restaurant in Abilene, Texas; Naugahyde, its maker, was showing it off in a salesman's looseleaf binder thick with product samples. "A startling 'furred' texture combined with authentic pattern reproduction" is how they trumpeted it—think herds of Indian horses on the wide Western plain—its plastic resins "expertly blended and permanently fused," into hard, hairy swirls of solid vinyl.

Naugahyde—the quintessential vinyl-coated fabric, icon of 1950s culture, to some a symbol of all that's cheap and cheesy, to many the stuff of bad jokes and knowing looks—is, of course, a brand name. For decades it marked products of the U.S. Rubber Company made in factories in Mishawaka, Indiana, and elsewhere. U.S. Rubber became Uniroyal, Uniroyal disappeared, and Naugahyde spun off on its own. A few years ago someone apparently tried to buy the storied name for some millions of dollars and outsource its production overseas. But it didn't happen and now the company still peddles Naugahyde made in its one factory in

Stoughton, Michigan. Almost identical stuff spews out from dozens of competitors, destined for the upholstery of your recliner, the cover of your checkbook holder, the seats of your outboard. But just as every photocopy is a Xerox, every tissue a Kleenex, it's all routinely, if wrongly, lumped together as "Naugahyde."

Ed Nassimi sells his company's own lines of vinyl-coated synthetic leather, bearing names like IMPERIAL, SYMPHONY, and TEAM PERFORMANCE. Nassimi grew up in Germany. His parents got into the business there in the 1970s, buying up the end lots of *kunstleder*, imitation leather, that became available at year's end when state companies needed to unload excess stock. Ultimately, the younger Nassimi moved to the States and set up shop as a kind of middleman for the German state economy. Today, with links to both Europe and Asia, the Nassimi Corporation maintains offices on the seventeenth floor of a building across West 31st Street from Madison Square Garden in New York City. Its boardroom chairs are covered in high-tech iridescent vinyl.

One Nassimi product, ESPRIT, comes in dozens of colors, like Regimental Blue and Sangria, and competes, says Nassimi, with a Naugahyde line introduced at the time of the American Bicentennial; this is Nassimi's Chevy or Ford, its generic imitation leather. Its "Symphony Collection of Faux Leathers and Suedes," on the other hand, offers numerous variations on a theme, many developed in the great vinyl-making factories of Taiwan; over the past quarter-century, he says, Taiwan has largely taken over from Germany as his prime source. One variation is impressed with a light grain reminiscent of kid or calf and offers "the elegant look and feel of soft glove leather." Another, which comes in colors like English Toffee, Ale, and Old Bronze, is treated so that if you tuft, stitch, or stretch it, a second color pops to the surface, imitating "the elegant look and feel of waxy pull-up leather," a fixture, presumably, of English clubs and boardrooms.

It's all vinyl, of course.

Poor vinyl. Cheap vinyl. Tacky vinyl. It stains. It tears. It rips. Doesn't just rip, actually, but rips raggedly, *unbeautifully*. Yes, as in the mantra chanted on behalf of plastics since Day One, it *cleans with a damp cloth*, making it ideal for, say, exercise equipment. But enough hair oil from enough sweaty athletes' brows and the plasticizer leaches out, leaving it brittle, prone to unsightly cracking. You've seen this; we've all seen this.

"It was like leather, but plasticky," recalled one consumer interviewed for a television segment about Naugahyde. "In summer," said another, "it would feel hot and sticky and make obnoxious sounds as you got up." Someone pictured it as a relic of Grandma's favorite couch, "almost too durable," its tacky vinylicity with you forever. Do Naugahyde and its vinyl brethren, I wonder, suffer doubly in the public mind, by conflation of two meanings of the word *tacky*? Tacky means sticky. But also, through a different etymology, shoddy and unfashionable.

Naugahyde, and its ilk, is all around us, part of our lives, used more than the leather it imitates. And it has, let us say, *points in its favor*, which those who make and sell it will be happy to list for you. It's mildew resistant and oil resistant, the spec sheet for one vinyl reports. It passes the flammability codes of the Boston Fire Code and of Federal Specification A-A 2950-A. It's formulated to block the growth of *Salmonella choleraesuis*, *Proteus vulgaris*, and a host of other microorganisms. You can get it in Space Blue, Taupe, Mandarin Orange, or pretty much any color you like so long as you want 1,500 yards of it. And it lasts and lasts; it will endure 500,000 cycles on the Wyzenbeek test, which tortures test samples with abrasives.

These and other vinyl virtues may not be those you most highly esteem, and they won't seduce a handbag designer for Gucci, but out in the market they count for something. Count, in part, because they complement another formidable virtue: Vinyl is cheap. Five bucks a yard? That's about what Naugahyde was getting in 2003 for a regular run of 1,500 yards or so; of course, it runs higher for short runs or special treatments. But, says Michael Copeland, a former director of styling for the company, "You can find $1.95 vinyl out there, too"—this when cheap grades of leather might work out to the equivalent of $20 a yard. Upholster seats with it for casinos and nightclubs, says Ed Nassimi and, day after day, you get guys slamming their thick bodies down into them, sweating away with nervous energy and excitement, wanting nothing so much as a little coddling, a hint of luxury. Vinyl, he says, supplies that. It sheathes your restaurant menu. It covers your basement recliner. And if its color and texture are done right, it makes some small, if not especially discriminating, corner of your brain think *leather.*

PVC is the polymerized form of vinyl chloride, a colorless gas with a sweet, mild odor known since its discovery in 1834 by Henri-Victor Regnault. Forty years later a test tube of vinyl chloride exposed to light seeped a white deposit, which we now know to be polyvinyl chloride, a hard, horny material that, when heated, gives off what one account describes as "choking and corrosive clouds of hydrochloric acid." Here was a material without a purpose. "People thought of it as worthless," Waldo Lonsbury Semon recalled shortly before his death, at the age of 100, in an Ohio nursing home. "They'd throw it in the trash." He was remembering the 1930s, when he helped create an industry.

B. F. Goodrich, the big rubber company, had in 1926 hired Semon out of the University of Washington to develop an adhesive that could bond rubber to metal, to better line pipes or tanks. Semon tried a variety of synthetic rubbers, got nowhere, then figured he'd try some simple organic polymers; one of several candidates was polymerized vinyl chloride. He prepared a batch, noted that at room temperature it didn't much dissolve in anything but that, heated in dibutyl phthalate and tricresyl phosphate, it did. And—here was the significant thing—when the mixture cooled, Semon found it wasn't hard anymore but soft, elastic, even jelly-like. The solvents formed a plasticizer, which altered PVC's properties enough to make it a new and useful material. Semon insulated the handle of a screwdriver with the stuff. He tried molding shoe heels and golf balls with it.

At one point, the way Semon told the story later, his wife was making shower curtains from heavy cotton. He took one, coated it with his new material, brought it round to the office of the vice-president, declared it waterproof, draped it over the veep's in-basket, and, before objection could be voiced, poured water over it. Needless to say, the papers emerged dry; in these stories of invention nostalgically recalled, the big demo always works perfectly. Later, when Goodrich posed Semon for a publicity photo, you see him, dark hair aglow in the dramatic side lighting, lab notebook open by his side, pencil at hand, decanting some chemical into a flask. It's the cleanest lab bench you ever did see, Semon's lab coat the best pressed and fitted. It was 1937, more than a decade after the discovery that left Semon, at 39, head of Goodrich's synthetic rubber program; his key patent, which would land him in the Inventor's Hall of Fame, had been awarded five years earlier. Plasticized PVC was big and getting bigger. English and

German companies projected huge new markets for it. At Goodrich early PVC production went toward shock absorber seals, coatings for electroplating racks, and coated fabrics like those we know today as Naugahyde.

Naugahyde itself goes back to Naugatuck, Connecticut, birthplace of the U.S. rubber industry, where some time before World War I, U.S. Rubber began using the name for rubber-coated fabrics, a nod to both the town of its birth and, apparently, the material it sought to emulate; aimed mostly at the automotive market, it doubtless competed with Fabrikoid. For years they made it like that, rubber-coated fabrics produced by U.S. Rubber were widely used during World War II in, for example, lifeboats.

Naugahyde didn't become a cultural icon until after the war; that's when rubber gave way to vinyl. A U.S. Rubber company history published in 1948 illustrates the change in a single pair of photos. One shows two horse-mounted policemen wearing bright capes, for visibility and rain protection, made from yellow rubber. Beneath it appears another twosome: Two little boys sit side by side on a bus seat wearing shorts, cute hats, and adorable smiles. Both have pulled their little legs and polished dress shoes, all rough soles and hard heels, onto the seat, making the parent in you cringe for the sake of the upholstery getting beaten up. Except it's not getting beaten up, the photo says, because it's "Naugahyde plastic upholstery," by now a key company product.

In the years after World War II, U.S. Rubber wasn't shy about describing Naugahyde as plastic. Vinyl, a company brochure told readers, "has many of the characteristics of rubber, but it is a PLASTIC," all caps; this could still be said without defensiveness. The brochure, all postwar modernity, shows a horse-and-buggy yielding to cars and buses. Plastics, say its authors, "are modern developments—the results of scientific research for better things," vinyl upholstery being one of them. One by one the virtues of plastic are ticked off: "Complete range of colors . . . not porous . . . color an integral part of the plastic . . . not affected by heat, cold, rain, fog, salt spray or sun exposure . . . not harmed by spilled foods or drinks, oils, greases." Naugahyde didn't embarrass the company; they liked to show it off. One *House and Garden* editor, apparently on something like a grand tour of U.S. Rubber, got a graphic demonstration, all big round dials and loud graphics, of Naugahyde defeating leather in side-by-side lab tests.

Bob Young, for years a Naugahyde man, started back in the 1950s as a clerk in a midsized Philadelphia company, Masland Duraleather. It was later bought by U.S. Rubber, which moved him out to Mishawaka in 1978. "Red brick, one gigantic square box with a lot of windows" and dried up wooden floors. That's how Young recalls the five-story, turn-of-the-century building, on the banks of the St. Joseph River outside South Bend, where he worked. On the first floor of the ramshackle structure, where the original Keds sneakers once were made, they manufactured Naugahyde.

The factory wasn't so much smelly, he remembers, as noisy. It was a place where forklifts banged and clanged down the aisles, hefting rolls of unfinished Naugahyde, bearing them from one operation to the next; hissings of steam, the rhythmic thrum of machines, great steel rollers turning, blowers roaring. Trays with printing inks, or topcoats, had to be refilled frequently, the inks pumped in from 55-gallon tanks. "And if you've got four pumps going simultaneously," Young remembers, "all going at different speeds, different things moving, everything playing its own little song, well, it's not earth-shattering, but it's LOUD."

There are several ways to make vinyl-based imitation leather. One is called casting, or coating, which uses a special "release paper" already lightly embossed with a leather grain. As the paper moves through the long machine, a hot vinyl syrup is fed into a trough between rollers and from there onto the paper—an advancing front of pure, perfect color spreads over the moving paper; baked as it passes through an enclosed oven, this film of vinyl becomes the top layer. Then, a second material is laid down on top of that, also vinyl but including "blowing" agents so formulated that when heated it rises like bread dough, forming a layer of foam. What's happening in stages across the line of rollers doesn't make intuitive sense right away, until you realize that this wide ribbon of material is upside down, its "flesh" side on top. Finally, paper and vinyl laminate split off from one another. The newly "cast" vinyl is wound onto rolls for later mating with a cloth underlayer. The paper is saved for reuse.

That's one way to do it. In an older, still frequently used method, you mix PVC resin, which comes as pellets or chips, with plasticizer and pigment, both thick liquids. It all goes into a big hopper from which it's deposited along a conveyor belt as a material with the consistency of peanut butter. From there it's into the reservoir of the calendering rollers, the heavy steel cylinders that squeeze it under heat: In goes the peanut butter, out comes a solid sheet of colored vinyl. The bare vinyl is then laminated to an underlying fabric in a huge oven—"like two railroad box cars end to end" is how Young remembers the one at Mishawaka. The cloth backing can be woven, nonwoven, or knit: Do you want a material you can pull around the corners of a sofa cushion, say, or one that feels softer to the touch, or some other mix of qualities? In any case, out it comes, wound onto 300-hundred yard rolls, then loaded onto a dolly for the next step.

And most always there *is* a next step. The art of making Naugahyde and its cousins is one of surface effects meticulously controlled, the vinyl-on-fabric laminate being printed with patterns, embossed, and finished off with this or that top coat. View Naugahyde sample books from the 1960s and 1970s, for example, and the range is extraordinary, with eFFect superimposed on **eff***ect* to create sometimes bizarre **E***ffe*CTS. BANGKOK suggested "fine Oriental silk enhanced with a subtle moire." WATERCOLOR was supposed to resemble "paint in a liquid dispersion." Most consistently emulated, though, was leather itself: "We looked over our shoulder at leather," recalls Young. "We called ourselves artificial leather. We tried to be like leather." Naugahyde STATUS boasted "a look and feel that suggests finely oiled doe's skin." In GREENBRIER a special printing and embossing process simulated stitching and a two-tone stain, giving "the outstanding appearance of hand-tooled leather."

It did nothing of the kind, of course; actually, it looked pretty sleazy. Still, each product demanded its own special effects, treatments, and production-line tweaking. Run the embossing rollers a little hotter? Reconstitute the printing inks? Reduce the pressure exerted by the cylinders? It was hard to get things just right. From time to time, Young remembers, you'd stop the line and cut out a big square of vinyl, about the size of a television screen—which gave its name to samples: "I've got a TV here," a workman would announce as he brought one over for inspection. There, under retina-dazzling light, they'd compare it with the color and grain it was supposed to have. Maybe it was "Old Gold" they sought, color EP-43, and they were getting too much yellow in it. Or maybe the color was okay, but it was going on too light. All this before they could comfortably settle into a 5,000-yard run of ENGLISH PUB, with its "'gentlemanly' burnished finish look."

None of this was rocket science? No, but back in the 1950s, for instance, U.S. Rubber had a substantial staff of chemists, physicists, and engineers at its Mishawaka plant. And its own Coated Fabrics Laboratory was forever trying out new treatments and effects. One was for a suedelike finish. According to a patent filed in 1958 the idea was to sprinkle fine, water-soluble granules onto the surface, raise the temperature so that they sank into the softened vinyl, then wash them away, leaving a surface so finely pitted it felt napped. Another patent protected a means of applying a "slip finish" consisting of silicone rubber dissolved in an organic solvent; the goal was to reduce the stickiness for which Naugahyde was notorious.

These, of course, were merely refinements superimposed on what, even by the 1950s, was a well-established technology—the exceptions, the little production line departures, that proved the rule. And the "rule," applied to most Naugahyde, as for most imitation leather? Well, it was to expropriate leather's "logo"—to borrow a particular pattern of wrinkles, veins, and haircells from a piece of real leather and stamp it into plastic.

I've brought with me to Richmond, Virginia, a vinyl notebook cover, in butterscotch and brown, that impresses me for the uncommon richness of its grain. This in a notebook costing $6.99, mind you, from the local drugstore. What, I wonder, can they tell me about it here?

I am visiting Standex Engraving, which makes the big embossing rolls that make sheet metal look like wood and vinyl look like leather. Gerd Mirtschin, supervisor of hand engraving at Standex, scrutinizes it now. His eyes, hands, and brain have been immersed in the intricacies of leather grain for decades. Though 48 years past his apprenticeship in Germany and near retirement, he is slim, boyish, still ardent about his craft. "Oh," he says finally, "*we* did that."

Excuse me? He's examined a few square inches of leather grain, one of thousands of grains impressed into the fake leathers of the world and that to the rest of us all look pretty much alike, and *recognized* it?

He reaches across to the side of his bench, riffles through a thick bundle of samples from past projects, pulls one out from halfway down, and silently turns it this way and that, upside down and right side up. "Yes, here it is," he pronounces at last in a light German accent. He shows me a section of the sample, points to an area of my notebook cover, and, sure enough, every irregularity, every wrinkle, every pimple and pore is identical. "I know every grain I did," he tells me later. "I just look at it and I know."

Standex issues a glossy brochure that shows its crafts people ministering to etched rollers. "Texturization" it's entitled, and across all six pages of text and color photos runs an inch-wide textured band—a vaguely Escher-like basketweave that feels like grapefruit skin. Embossed surfaces like this one, of course, are everywhere—on paper towels, book covers, vinyl flooring, paper doilies, faux wood. And somehow we know, or think we know, how they got there: Why, you press your thumb into cookie dough, lift it, and your thumbprint remains; what's to say? What more needs to be explained? An encyclopedia account devoted to "Leather, artificial," from around 1910, advises that leather grain is imparted to the material "by passing it between suitably embossed rollers." No word, though, about just *how*, "suitably" or otherwise, those rollers get embossed.

In the beginning, plainly, comes the hide, a piece of real leather with variegations of grain, swirls, sweeps of texture, and charming little blemishes deemed worthy of emulation.

In the end we find its pattern represented on embossing plate or roller.

But in between? The years have yielded many methods, the surprising twists of the problem inspiring the care and attention, not to say genius, of artisans and engineers around the world. They've used gouges and chisels to replicate leather by eye. They've covered leather with conducting mixtures and electroplated it. At Standex old and new come together in great steel cylinders, metal-eating acids, greasy inks, molettes, chisels, silicone molds, computers, and room-sized lasers.

In one corner of the plant, a laser reads the undulations of a leather grain, much as it might the microscopic pits of a compact disk. The three-dimensional information it stores is later used to control a second, more powerful laser, which burns into a rubber roller, its surface left resembling the original leather. That roller, made to bear up against a vastly larger steel roller coated with protective ink, lifts off some of the ink. Where it does so, an acid bath, in successive dippings, can attack it, etching the original pattern into the steel. This steel roller, now "suitably" etched, is ready to emboss the vinyl. And that, in crude outline, is that.

Of course, this herky-jerky Rube Goldberg-like train of mechanical logic obscures all the practical difficulties. The original grain passes back and forth between positive and negative, through different materials, several times. Numerous mechanical problems must be surmounted to repeat the pattern, step it across the full width of the embossing roller, and keep everything in proper register.

But trickier by far is this: That first laser scans the leather in three dimensions, recording the depth of each swell, swirl, and pockmark, right? And all those points, taken together, make for a topographic map of the leather's grain, right? Well, no, not quite. Because to invoke a precision that now becomes necessary, the topographic map is just of the *sample* grain, the little piece of leather being copied. A cowhide might cover 50 square feet, a select sample of it perhaps 1 square foot. And the vinyl shipped to the customer? Maybe 4 feet wide by 2 *miles* long. And in those seemingly innocuous numbers lurks a problem, as applicable to Standex's laser technology as to all the methods used over all the years to make embossing rolls and plates: How, exactly, does one small sample of real leather grain show up on that much larger piece of faux leather?

Well, the pattern simply repeats, of course. The hot roller turns, bears down on the vinyl, impressing it with the grain of the original. The roller continues to turn, with each revolution bearing down on a new stretch of vinyl, thus impressing *it* with the grain of the original. And so on, at maybe 20 yards per minute, the sample endlessly repeated.

But what of the border between adjacent sections?

Ignore that question and your 2 miles of imitation leather would be punctuated every foot or two with an ugly, awkward edge, corresponding to each revolution of the roll: a nice leathery region, then its abrupt end; another nice leathery region, another abrupt end. Indeed, that's just what sometimes happened with early plate embossing methods which, as one technical look back recounts, "were unsatisfactory at the edge of the embossed pattern where succeeding plate impressions formed a skip or overlap in the continuity of pattern." At Du Pont, in 1915, Pierre du Pont himself could be heard to complain that one expanse of Fabrikoid he'd seen looked like a checkerboard, the patch of grain pattern repeated too conspicuously. Ideally, then, the faux leather grain must *never* simply end but must metamorphose into a pattern without end or beginning.

In the Standex laser department they handle the edges by airbrushing them away. Literally. They eliminate the edges on the digital image of the grain, before they can be embodied in steel, rubber, or other intermediary material. Tommy Austin, laser engineer, uses an ordinary computer mouse to highlight a border region on the screen, digitally lays down over it a new section of grain borrowed from elsewhere in the sample, then uses computer tools with names like AIRBRUSH and CLONE to soften and blur the juncture; were you now to fold top and bottom of the pattern into a cylinder it would, where the two edges meet, flow seamlessly into one.

The digital information recorded controls a kind of laser lathe. Pulsing to the tune of the pattern's digital record, this powerful laser burns into the spinning hard-rubber roller. What remains on its surface, once you wipe away the fine cloud of rubber dust left behind, is, more or less, the leather grain's original three-dimensional pattern. "More or less" because its edges have been altered.

At the other end of the Standex property, they still use fine chisels to smooth over the awkward edges that Tommy Austin manages digitally. Before becoming a laser engraver, Austin had worked for 20 years as a hand engraver. Gerd Mirtschin still does, working in a brightly lit room with no lasers and no computers; he starts with a piece of leather whose best region to copy he sometimes gets to select himself.

Mirtschin begins by taking a silicone impression of the leather sample, which thus represents a negative of the original. From it, a positive is made out of epoxy, which is then wrapped around a steel roller. Then, in the crucial, painstaking step, this epoxy-sheathed roller transfers its intricate surface detail, traceable to living skin, to a small steel die.

The die starts out as a plain steel cylinder, perhaps 3 inches in diameter and 6 inches across, completely covered with a greasy, protective ink, an "acid resist"; were it now immersed in an acid bath it would come away unscathed. But remove any of that ink and the acid could start eating it away. Remove which ink? Remove only that ink which corresponds to grain markings in the leather. Well, roll that epoxy-surfaced roller bearing the grain over this cylindrical die, and it will pull up ink at just those places corresponding to the "top" of the grain. If you now immerse the die in acid, the first faint traces of grain pattern will appear; a valley from the leather grain shows up as the slightest of ridges.

Now, repeat the process: Re-ink the die, lift off a slightly "lower" bit of grain detail, and etch for a few minutes, the pattern reaching down just a little further into the surface. Do this perhaps 100 times—the slightest bits of metal removed each time, three-dimensional detail revealed with ever greater precision—and in the end that one small die will carry virtually all the detail of the original piece of leather.

But Mirtschin faces the same problem encountered by Tommy Austin in the laser department: The pattern represented on the steel die must be multiply copied onto the embossing roll in such a way that the vinyl bears the illusion of a single uninterrupted piece of leather. "You have to work the joining out," says Mirtschin. "Nature doesn't have that." And to do that he uses not a computer mouse but chisels—slim toothpicks of steel a few inches long, their tips fashioned into points, lines, and tiny scoops, gathered into groups on his workbench, hundreds of them. With them he "makes" haircells, pebbled grain, wrinkles, creases, and pores. The handle of his hammer is made of oak he's shaped to his own hand and flows so gracefully as to qualify as sculpture. With it he deftly swipes, chips, and cuts away metal surfaces so that, by the time he's finished, every awkward, man-made-looking edge is gone from the embossing roll, leaving a single, free-flowing, seemingly organic grain.

I ask Mirtschin to demonstrate, but he refuses. Oh, in the end he hauls out some chisels and takes a few desultory swipes on a piece of scrap metal to show how he does it. But he politely declines to demonstrate on his current job. Sorry, he explains, you don't just pound away; you need to study the flow of the grain. That's how he recognized my notebook cover.

Four hundred miles away, in a vinyl-making plant in Ohio, stand racks of embossing rollers, some of them from Standex. They are lined up in rows, separated by aisles, hundreds of them; many more are in storage. Not all bear leather's imprint. Some carry more abstract or stylized patterns. Some are used for printing or for applying adhesives. But many do carry a leather grain. Look at any one of them and you see it plainly—leather's logo, only in reverse, fine furrows in the hide here represented by low mountainous ridges. Run your finger across it and it feels a little abrasive, like worn sandpaper.

We're in the SanduskyAthol plant, in Sandusky, Ohio, 50 miles west of Cleveland, a Naugahyde competitor. Bales and rolls of material in various states of completion. Pipes and pumps and rollers. One hundred forty people in this cavernous place calender vinyl or cast it on release paper. Virtually none of it leaves as bare untextured vinyl. One way or the other, most of it is embossed, the vinyl passing between cylinders heated to 375 degrees Fahrenheit, then between chilled rollers, water droplets condensing on their cold metallic surfaces.

The embossing rollers, I'm told, cost upward of $30,000 each, but there's no reason they shouldn't last forever, if they're taken care of. And they are. Sitting heavy and immobile on their racks, they're swaddled in special cushioning held tight by velcro closures to make sure every square inch is protected. A yellow tag affixed to one warns:

CAUTION
PRECISION ENGRAVED ROLLER
THE SLIGHTEST DAMAGE MAKES IT USELESS

The moment one of these great rollers bears down onto that wide river of vinyl is the moment faux leather becomes itself. It, or something like it, has been used to emboss Leatherette, Fabrikoid, Naugahyde. With it a material that is not leather is made to look like leather.

With it a fraud is perpetrated on the world.

With it a moment of high industrial artistry is realized.

World War II was over, and in Newburgh, New York, where they'd been making pyroxilin-based Fabrikoid for half a century, they were making mostly vinyl now; for a while, remembers Sam Lange, who started with Fabrikoid in 1951 as a process engineer, they made Fabrikoid in one building, vinyl next door. By the time he left in 1967 it was mostly vinyl. Pyroxilin had lost the automotive business, Lange recalls—except, as it happens, for that little shelf under the window behind the rear seat. Because there, still bearing its requisite leather grain, it didn't have to flex.

Vinyl stood up better, especially where you encountered flex, as in seats. Soon after the end of the war, the London Passenger Transport Board chose it to replace the Fabrikoid-like material long in use. Around this time, Storey's of Lancaster largely switched over, too, importing vinyl production machinery from the United States. "The year 1951," a company history records, "found the firm launched upon a sea of plastic." By 1967 the United States was producing 142 million yards per year of vinyl-coated fabrics, versus 19 million yards of its increasingly obsolete predecessor. In these postwar years, if you wanted to make something serve for leather, you turned to vinyl.

Among vinyl makers, it was Naugahyde, of course, that became the emblem of 1950s culture. U.S. Rubber hitched it to the do-it-yourself craze, a company publication showing, step by step, how to reupholster battered old chairs with it. Buckminster Fuller used Naugahyde in his igloo-shaped Dymaxion house of 1946, green in the bedrooms, butterscotch in the living room. Naugahyde covered chairs at the United Nations building in New York. It was everywhere.

These days you sometimes hear fond reminiscences of "genuine imitation Naugahyde." Grace Jeffers, a New York design consultant and historian of faux materials—the daughter of a Formica salesman, she wrote her master's thesis, "Machine Made Natural," about Formica—still associates Naugahyde with her 1970s Midwest childhood. For her, Naugahyde meant "the whole family getting together. It's always in places where people gather, like a breakfast nook, always associated with intimate, domestic moments." Naugahyde was a delight. "It has a candylike quality to it. It's fun and easy. Think of where you see it, like in an amusement part, or the seats of a diner."

Of course, a lot of the would-be Naugahyde she and the rest of us encountered wasn't really Naugahyde. By the mid-1960s, Uniroyal had numerous look-alike competitors; the world's premier maker of imitation leather was itself being challenged by imitators. So Uniroyal approached George Lois for help.

Lois was the Madison Avenue advertising guru who'd helped sell the original Volkswagen to America. The way he told it later, Naugahyde was "a superb leather substitute" that had spurred imitators, inundating the market. Poor confused shoppers "could not distinguish marvelous Naugahyde from its many inferior imitations." To help them navigate this perilous terrain, Lois conceived the Nauga, a mythical animal that shed its hide once a year for the good of mankind.

Really.

Lois's Nauga had way too wide a mouth, way too pointy teeth, way too close-set eyes, something like an owl's ears, and a bulbous, uncool, unsexy body. It was comical, distinctive, and memorable. "The Nauga is ugly," said the ads, "but his vinyl hide is beautiful." They made hangtags featuring the Nauga. They made children's toys. A Nauga once showed up on the *Tonight Show* opposite Johnny Carson. In later years the Nauga became something of a cult phenomenon. They were made available for adoption. Nauga dolls were sold on eBay; one collector wound up with 60 of them and posed for a *New York Times Magazine* article with her display. Even today Naugahyde carries "A Nauga Story" on its website, with eight pages of loony cartoon images and absurdist history. Naugas inhabit a Pacific atoll. They immigrate to Ellis Island, their line extending to that eminent robber baron, Cornelius VanderNauga, pictured on an overstuffed, presumably Naugahyde-upholstered chair in his Newport, Rhode Island, mansion. None of it is terribly clever, being all the expression of a single, slight idea. But mention Naugahyde today and it's not long before you're being knowingly reminded of "those cute little Naugas." These days, the fiction is maintained; they live on a secret ranch near Stoughton, Wisconsin, where, as it happens, Naugahyde is manufactured.

Whatever good the Nauga may have done for the company's bottom line, it did little to alter the perception of Naugahyde as incarnation of all that's inauthentic and unreal. "Is it real or is it Naugahyde?" was the refrain of a five-minute television segment devoted to it on one of those *Whatever happened to . . .?* nostalgia programs. Naugahyde was a fixture of what design critic Thomas Hine has called the Populuxe period, from the mid-1950s to the mid-1960s, which bestowed on American life a peculiar, yet enduring, species of popular luxury built on an edifice of FAUX. Populuxe was built on Formica, which drove out the enamel-metaled kitchen. And Con-Tact paper, an adhesive-backed vinyl film that could be printed with stripes or squares, marble or wood, and that, writes Hine, "turned anything into something else entirely." And of course, Naugahyde itself, which became an enduring symbol of the falsity of modern life.

Just mention Naugahyde and it's as if you change the subject on the spot, are propelled into the realm of the synthetic, the inauthentic, and the surreal. A *Time* book review described a Rick Moody novel, *The Ice Storm*, in which old-line Connecticut families "have exchanged Chippendale propriety for Naugahyde and wife swapping." On the Web, someone asking what Naugahyde was got the straightforward reply: "It is to leather what Formica is to marble." An author responding to a negative review of his book about intelligent design asserted that his critic "inhabits a fantasy world populated by a fantasy life that has no more connection to biological reality than Naugahyde has to cowhide." Others see fraudulence embodied in its very name, as "a way to fool the unwary," in the words of one

Web correspondent, "into thinking that the material was actually the hide of something that once lived."

Today the Cooper-Hewitt design museum, housed in the old Andrew Carnegie mansion in New York, keeps samples, catalogs, and brochures from every design epoch, style, and taste, from Naugahyde to Gucci. Stunning as they are, the Gucci catalogs seem ephemeral, reflecting the shoe and handbag fashions of particular times. But Naugahyde brochures from 1964, for example, illustrate a world that for the most part is still with us, a semipublic sphere of restaurants and hotels intended to suggest English club rooms of overstuffed, button-tufted opulence—but which somehow reek of unreality. Are one brochure's wood-paneled walls really wood? Are the plants alive? Are the linen-look tablecloths really linen? Is the marble-topped table really topped with marble? Certainly, these tableaus are intended to evoke marble, wood, linen, and leather. But our doubts as we step through these Naugahyde-inspired visions of modern life—that faint, inexpressible uncertainty, that unease as we sense the split between look and feel, appearance and reality—seems borne along on the soft, vinyl shoulders of Naugahyde.

Vinyl can seem at one additional remove from nature than earlier faux leathers. Leatherette, after all, was made from felted paper, Fabrikoid from cellulose; ultimately, both came from plants. The ethylene and saltwater going into the forty billion pounds of PVC manufactured each year, though, never lived. The Vinyl Institute tries to humanize it, pictures it as the most versatile plastic, readily recycled, most of it derived from common salt, not all that poisonous. Still, made as it is in sprawling chemical plants from petroleum, broken down into ethylene, and thence into vinyl chloride, totally synthetic, mixed with pigments and plasticizers, it's not hard to imagine why in Naugahyde and its ilk we might find something more troublingly alien, more oily and unfamiliar, than any of its faux leather predecessors.

But in fact, viewed just a little differently, plastics like PVC and natural substances like collagen are sisters under the skin, not so much different as the same.

5

One Nature

In the fall of 1927, Wallace Hume Carothers, a 31-year-old organic chemist, was lured from Harvard University by Du Pont. Carothers, product of a tiny Presbyterian college in the American Midwest, had gone on to get a Ph.D. from the University of Illinois and was, at the time Du Pont found him, teaching chemistry at Harvard. There he impressed everyone with his brilliance. Word got out, Du Pont made him an offer, and ultimately won him over with the promise of research freedom, able colleagues, the best equipment, and limitless funds. "A week of industrial slavery has already elapsed without breaking my proud spirit," he wrote early the following year.

Carothers was sensitive and shy but thought big, aiming to synthesize the sorts of big molecules—polymers—everyone was talking about in those days. This he did; nylon, which showered untold riches on Du Pont and became an indispensable part of modern life, was one of them. In Carothers's short, haunted life—he killed himself in a Philadelphia hotel room at the age of 42 by swallowing a cyanide capsule—he became something like the 20th-century's chemical counterpart to Thomas Edison, his lab a prodigious innovator of new materials. But it also contributed to a larger scientific debate about the nature of the polymers he and his colleagues were making.

A polymer is just a compound made up of many smaller molecular subunits—a "monomer"—repeated. In the case of nylon, the monomer includes an amide group, a particular arrangement of carbon, oxygen, nitrogen, and hydrogen; nylon, then, is a polyamide, and synthetic.

But polymer scientists were just as interested in substances derived from nature, like rubber. Natural rubber comes from trees, whereas nylon comes from a factory, from *chemicals*. But the distinction means little; rubber is a polymer, too. Other natural polymers include cellulose, gelatin, and its precursor, collagen. Nylon's emergence as part of the everyday material vocabulary of our world was owed, in part, to fruitful work in the laboratory. But it was owed, too, to more than a century of scientific effort that helped cut across the old, seemingly impermeable, boundaries between animal skin and test tube, tree and factory, "natural" and not. During the 1920s and 1930s, chemists like Carothers worked with synthetic polymers to learn about natural polymers and with natural polymers to help them make synthetic ones, distinctions between them getting scant attention compared to all they held in common. In 1942 appeared the English translation of a classic German polymer text. In his introduction, author Kurt H. Meyer proposed to classify organic polymers by chemical composition and so "ignore the division between natural and synthetic compounds, a division which is no longer made in organic chemistry."

What distinguished all polymers, synthetic and natural, from most other substances was that they were *big*. That is, the molecules making them up were big. Whereas sugar, salt, common acids, and most other denizens of the chemistry lab passed readily through certain familiar membranes, these materials did not. The molecules constituting them, plainly enough, were large, hindering their passage through the membranes. How large? And large in what way? On these questions, well into the 1930s, opinion sharply differed.

The most prevalent view, at first, was that they were small molecules held together in larger aggregates by electrical or other forces no one ever quite defined. The big unruly substances that couldn't pass through membranes were termed "colloids." The elements, or subelements, or whatever they were that made them up were called "micelles." What governed their behavior were principles no one understood but were assumed to differ from those of more familiar chemical compounds.

But this view wasn't shared by Hermann Staudinger, an organic chemist at the Institute of Chemistry in Karlsruhe, Germany, who, beginning about 1920, advanced quite a contrary notion. What made rubber, say, or collagen's cousin gelatin, unable to pass through membranes wasn't that they were modestly scaled entities mysteriously bound together by unknown and unexplained forces. Rather, they were huge, otherwise ordinary molecules held together by the same subatomic forces that held every other molecule together; in Staudinger's language, they were macromolecules. The molecular weight of carbon was 12, that of carbon dioxide about 44. Simple organic compounds such as gasoline or sugar had molecular weights below 500. A colleague of Staudinger once assured him that molecular weights beyond 5,000 were impossible. No, Staudinger stoutly insisted, they could run into the tens of thousands, hundreds of thousands, millions. A macromolecule was not, by crudest of analogies, like an assemblage of a toddler's 10-piece jigsaw puzzle uncertainly yoked together with string and paper clips. But rather, like a single room-sized Ultimate Jigsaw Puzzle, with millions of pieces, all intricately fitted together—just like any other puzzle, only bigger.

For 20 years the idea that would ultimately win Staudinger the Nobel Prize was just a theory, challenged at one scientific conference after another. At a meeting of the Zurich Chemical Society in 1925, many tried to convince him of how wrong he was. "The stormy meeting ended," one observer reported, "with Staudinger's cry [mimicking Martin Luther's], 'Here I stand, I cannot do otherwise!'" At a later meeting a chemist likened the natural skepticism toward macromolecules to that of a zoologist advised that "somewhere in Africa an elephant was found who was 1,500 feet long and 300 feet high."

But back in Delaware, Carothers had taken a fancy to Staudinger's ideas and set to making 1,500-foot-long elephants. He was taking monomers, those small repeat units, and chemically joining them; at the juncture between amides, say, he was whisking away hydrogen and hydroxide and condensing them into ordinary water; nylon, then, was dubbed a condensation polymer. And the chemical bonds uniting them? They were just like those of any other molecule. In Carothers's lab, many amides linked up to form a polyamide—nylon. In other labs, polymerized styrene got you polystyrene, and many vinyl chloride monomers locked in profound chemical embrace yielded polyvinyl chloride.

In his 1953 Nobel Prize speech summarizing scientific ground gained over the previous three decades, Staudinger placed macromolecular substances into three categories: First, wholly synthetic; second, natural, like rubber, proteins, or nucleic acids; and third, derived from natural substances, like vulcanized rubber, rayon, or cellophane. All were governed by identical principles, the substances Carothers and his colleagues made in the lab equally with those that humans had known since time immemorial, like collagen. Indeed, DNA, the material that transmits genetic information through generations of animals and plants, whose structure James Watson and Frances Crick had unraveled earlier in 1953, can be seen as a polymer built up from four nucleotide bases, their order specifying genes. *DNA? Nylon?* It wasn't two realms these materials inhabited but one.

One Nature, not two.

As it's dumped into hoppers that will ultimately go into a mix that becomes Naugahyde, PVC is a white powder, made in a chemical plant. Collagen comes from dead animals. One never appears freely in nature; the other is abundant. Yet, as we've seen, they are not so wildly different as, through a more cultural or social prism, they might seem. To reassert what should be plain by now, they are both "chemicals." Both are polymers, built up from smaller molecular units. Their behavior is constrained by the same chemical and physical laws. They are susceptible to similar tools of scientific analysis. They are part of the same Nature.

To Carothers, one of his biographers has written, "There was no breach between natural and artificial processes"; it was a largely meaningless distinction. Carothers wrote:

> The idea that natural high polymers involve some principle of molecular structure peculiar to themselves and not capable of being simulated by synthetic materials is too strongly suggestive of the vital hypothesis, which preceded the dawn of organic chemistry, to be seriously considered.

The "vital hypothesis," or "vitalism," went back to Aristotle and imagined a sharp boundary between living and nonliving, the former set apart by a kind of inspiriting, vitalizing force. In accord with this outlook was the division of chemical substances offered early in the 19th century by Swedish chemist Jöns Jacob Berzelius. Some substances, like olive oil or sugar, came from living organisms and thus, he said, were "organic." Others, like water or salt, came from the nonliving world; these were "inorganic." The two inhabited opposing realms, the line separating them an object of fascination and reflection to many but generally thought inviolable. What was Mary Shelley's 1818 novel, *Frankenstein*, but, among other things, a rumination on the border between the two and the fearful dangers of crossing it.

In 1828 a former student of Berzelius, 28-year-old German chemist Friedrich Wöhler, delivered a blow to vitalism. In the course of other work, he treated silver cyanate with ammonium chloride, yielding ammonium cyanate, a simple inorganic compound thought to have no special significance. But its colorless crystals, some of them an inch long, looked to Wöhler just like those of another substance, urea. Over the centuries getting urea meant getting it from the urine of animals, including man, thus making it, by Berzelius's scheme, indisputably organic. This ammonium cyanate, on the other hand, was the product of a reaction between two other inorganic compounds and so indisputably inorganic. Yet as Wöhler wrote in a paper, "On the Artificial Production of Urea" in the *Annalen der Physik und Chemie*, the two substances proved "absolutely identical." Materials devoid of life had, through routine chemical manipulation, yielded a substance fairly fragrant with life. "I must tell you," Wöhler wrote his esteemed teacher, "I can make urea without the use of kidneys, either man or dog."

That wasn't singlehandedly the end of vitalism. Chemists and historians of science have questioned the significance of what Wöhler did. And even today many, religious or not, conceive some sort of life force lying beyond the understanding of science. But surely the Wöhler synthesis, as it became known, blurred the once seemingly sharp division between the two realms. As historian of science Bernadette Bensaude-Vincent has written, it and later developments in organic chemistry showed "that what had seemed a permanent roadblock was instead a permeable membrane." Later, chemists synthesized a host of organic compounds, such as acetic acid, methane, benzene, and acetylene. And with the work of French chemist Marcelin Berthelot, it's been written, "Crossing the line from inorganic to organic ceased to be a thrilling intrusion upon the 'forbidden,' and became purely routine."

The early engine of organic chemistry was artificial dyes, made from coal tar, which came to largely replace dyes made, at great trouble and expense, from roots, insects, and snails. Rubber was another early target of the chemists. In the middle ages, alchemists had tried to make gold from base metals. Now, through organic chemistry, these new alchemists were synthesizing from cheap, widely available materials much-coveted substances that nature had seen fit to make in insufficient number or that required undue trouble and expense to extract. The powerful tools of organic chemistry let scientists synthesize substances almost routinely—in the quarter-century after 1919, one prominent chemist estimated, more than 200,000 of them.

"Until recently," said Charles Stine, the Du Pont chemist who had brought Carothers to Du Pont, in a post-nylon declaration that suggests the boundless confidence of the field in 1942, "man spent his efforts inventing superior uses for what nature had laid at hand. Sheep, plants and worms supplied fibers. Bones, tusks, horns, the saps and barks of trees, the excretions of insect and animal life filled trucks and freight cars and ship holds just as they had made up the caravans in the days of Marco Polo." But these weren't raw materials, not really; they were the finished products of another manufacturer, Nature. The real raw materials that humans had at their disposal "were a small group of chemical elements numbering less than a hundred. Carbon, hydrogen, nitrogen and oxygen were the true building blocks of nature."

Building blocks that formed vinyl and that formed collagen.

Run your fingers across the flesh side of a piece of leather and you're feeling mostly collagen; place it under a magnifying glass and you can easily see its collagen fibers. But those fibers, you'll see if you look more closely, are each bundles of thinner fibers. And *those* fibers are made up of still finer fibrils. Indeed, the structure of leather can be roughly likened to a fractal, which is a mathematical "object" that looks similar at every scale of magnification—like the coast of Maine, say, with its broad bays including within them baylike inlets, and these inlets in turn containing tiny baylike irregularities, and so on, each view looking much the same. Likewise for the collagen in leather which, as British chemist Edwin Haslam observed in a text devoted to vegetable tannins, has "a fibrous appearance at all levels of optical resolution available," down even beyond the visible range, in fact, to that of the scanning electron microscope.

The scanning electron microscope reveals startlingly clear striations along the length of each collagen fibril—bands or lines separated by about

67 nanometers, or billionths of a meter; each marks the regular staggering of individual molecules lined up in parallel array. If you could peer still more deeply into the molecules themselves—you can't, because you're beyond anything you can call "seeing"; the rest is inference—they would resolve into three interlinked helical strands wound around one another, a triple helix.

All proteins are built up from just 20 amino acids, attached end to end, like beads on a necklace. Their order—for example glycine, serine, glutamine, glycine, alanine—constitutes what molecular biologists call its primary structure. One sequence, for example, gives hemoglobin, which carries oxygen through the blood. A second gives insulin, which regulates sugar metabolism. A third yields collagen. Each "necklace," composed of a different number and assortment of those 20 different jewels, is different; each, one of the dizzying variety of permutations possible, makes one of the thousands of enzymes, hormones, tissues, and other proteins that together account for life.

The differing biological function of each results from its molecular shape, but this is normally nothing so simple as a necklace. The amino acids of which it's made pull and tug at each other. And that twists the superficially necklace-like molecule into convoluted shapes that molecular biologists routinely liken to accordion pleats, helical springs, and other familiar and not-so-familiar tangles. So it is with collagen, which takes the form of three helices, each of about 1,000 amino acids, all intertwined.

It was during the early years of molecular biology that chemists took their first shots at understanding collagen; no less a figure than Francis Crick himself, later of DNA double helix fame, was among those to suggest structural possibilities. By the 1960s the essential features were established. While broadly similar, collagens do vary a little by species, within species, and by the part of the body in which they are found. The kind found in animal skin, which typically forms 90 percent of the collagen in the mammalian body, comprises two identical helices and a third that's a bit different. But in all three, for most of their length, the tiny amino acid glycine appears like clockwork, every third jewel in the necklace; only so small an amino acid, it seems, can squeeze into the tight center of the triple-stranded helix.

Comfortable scientific certainty? Loose ends neatly tied up? In fact, mysteries abound. For example, cowhide is mostly collagen and, left untanned, rots. Leather is mostly collagen and doesn't. Just what, in tanning, spells the difference?

For a single word that might be guessed to represent a single thing, "tanning" means quite different things. Stick a pair of wet leather boots on a radiator, leave them there too long, and you may not be able to get into them once they dry out; leather shrinks, particular leathers at particular temperatures. And that temperature, which leather chemists accord its own symbol, T_s, is a sharp marker of its properties. Chrome-tanned leather has a markedly higher T_s than vegetable-tanned leather—110 degrees Celsius or so, compared to 85 degrees, just above the boiling point of water versus just below. This difference, coupled with those cited in an earlier chapter, most of them consistent and pronounced, tempts one to ask whether these two species of leathers are just minor variations on a theme or, somehow, fundamentally different. That is, if they differ so much, just what about them, if anything, is the same?

Moreover, vegetable and chrome are not the only ways to tan. American Indians would wrap the brains of an animal they'd killed in a cloth, boil it, knead the hide with it, then smoke the hide over a damp wood fire. Eskimos did much the same with reindeer hide. The skins of chamois, antelopes from mountainous regions of Europe, were tanned using cod or whale oil, yielding exceptionally soft leather. So it can seem little more than lexical accident that these different methods are anointed with the same term. But in fact they mostly earn it. Because however different, each takes animal skin across the boundary between impermanence and stability. Each results, more or less, in true leather. Each works in, with, or through collagen. It's just that after 5,000 or so years of tanning, it's still hard to pin down precisely how they do it.

"Tannage," declared Hubert Wachsmann, a former president of the International Union of Leather Technologists and Chemists in 2004, "has to change the properties of collagen, either by chemical reaction or by covering the fibers against outside influences." That's about the most unassailable thing you can say about tanning, a kind of conceptual bottom line: One way or another—and they differ depending on the kind of tannage—that dense matrix of collagen fibers must be rendered inert, invulnerable to those biological processes that would otherwise corrupt it. But while chemists in America, England, India, Germany, and elsewhere have advanced numerous theories for just how, at various levels of detail and sophistication, it's fair to say that none have proved in every way satisfying.

"Crosslinking" is the concept most frequently, if loosely, invoked among leather chemists. It refers to chemical linkages in tanned leather thought to make its molecular networks more stable than individual collagen molecules would be on their own. Yes, leather fibers are composed of fine fibrils. And yes, each fibril is composed of collagen. And collagen is composed of three intertwined helices. But that's not the end of the story. The triple helix has loose ends, or looser ends, anyway—areas of positive and negative charge, molecular corners and protuberances, that can link up with nearby others, perhaps just holding tight until something comes along to disrupt it, perhaps forming a virtually indissoluble bond. Tanning, the idea goes, promotes such crosslinking among collagen molecules, stabilizing them and rendering them invulnerable to bacterial attack.

Certainly, *something* like this is going on. In talking to laypeople, at least, Nick Cory, director of the Leather Research Laboratory in Cincinnati, pictures collagen locked in a tight protective cage formed by crosslinking forces. For Waldo Kallenberger, a chemist and microbiologist with a lively interest in every aspect of tanning and leather chemistry, crosslinking inhibits the enzyme action upon which bacteria rely to cut up their food, blocking putrefaction—but differently, depending on the tanning method. Chrome tanning interposes covalent bonds, classic crosslinks, between and among the chains. Vegetable tanning relies on weaker hydrogen bonds. "They just lie alongside the proteins and cuddle with them" is how he invites you to imagine them, "but they're pretty easy to wash away." Visualize, he says, chains, real chains, entwined in a pile on the floor. In vegetable tanning they're difficult enough to untangle just as they are. In chrome tanning the individual chains are tied together by supplementary links, making them that much harder to unravel.

Of course, all this carries a whiff of analogy and metaphor. But much real uncertainty remains. Back in the 1950s the Cooper Union Museum in New York held an exhibition devoted to leather; the author of its catalog explained that in tanning the idea is to "surround the individual fibers within the hide with some preservative substance that they become proof against bacteria or other decomposing agents." Half a century later, science's explanatory power far exceeds that—yet not quite so far as you might think. "In spite of all the extensive literature on tanning and tanning agents," two collagen scientists with the U.S. Department of Agriculture wrote in 1977, "serious gaps remain concerning the nature and extent of crosslinking and stabilization" involved in tanning. And gaps persist today. In 2001 the *Journal of the American Leather Chemists Association* carried a long (and unusually heartfelt) scientific essay, by Thirumalachari Ramasami of the Central Leather Research Institute in Madras, devoted to the author's dream of a "Unified Theory of Tanning." Any such theory, Ramasami concluded, remained elusive. Tanning, he wrote, "is more easily described than defined."

You can almost hear the intellectual frustration bubbling up from the page. Noting the substantial differences between various kinds of tanning, A. D. Covington, one of the world's foremost authorities on leather science, raised this seemingly straightforward yet maddening question in a lecture before Britain's Society of Leather Technologists and Chemists: "If the effects of tanning reactions are so varied, what is the definition of tanning?" What, precisely, *is* it?

Is it somehow appropriate that, whatever our faith in the power and exactitude of the scientific enterprise, a measure of mystery yet surrounds this ancient technological miracle and its modern counterparts? Tanning takes hide from an animal perhaps moments ago alive, still vulnerable to the life-and-death processes of the natural world, and carries it across a divide that leaves it largely immune to them. Of course, the mystery that yet surrounds tanning can be written off as nothing more than vestigial, any remaining dollops of uncertainty one day yielding, we may presume, to perfect understanding and precise technical control. But viewed alternatively, tanning can seem to reach between the natural and synthetic realms, link the living and the dead, operate at that uncertain, uncharted juncture, of such continuing interest to writers and philosophers, between the world built by nature and that contrived by man.

6

Nothing Like Leather

By the 1880s machines were already taking big bites of the world's work, but they ran not on electricity or gasoline but waterpower or steam. Large, centrally located engines transmitted power to individual machines, like lathes or looms, through long, snaking belts, looped over pulleys, whirring at three or four thousand feet per minute. Most of these belts were made of leather. Forests of them grew in every factory. They could be thirty or forty feet long, a foot or more wide, built up from individual hides spliced, glued, laced, or riveted together; at one point close to a hundred manufacturers made them. And they were important enough that, in 1884, Frederick Winslow Taylor began a study of them that would last nine years.

An engineer with a Philadelphia manufacturer of heavy railcar components, Taylor repeatedly saw great machine tools, the kind that cut, drilled, and shaped solid metal, reduced to impotence when a belt broke, or stretched so much it slipped off its pulley, bringing production to a halt. For Taylor, the first efficiency expert, that was the great sin. He began experimenting with his belts. He adjusted speeds and tensions. He tried belts made from leather tanned in varying ways. He recorded stretch, repair frequency, upkeep costs, and pretty much everything else. His results appeared in 1893, in a mechanical engineering journal, where they took up 189 pages.

Industrial strength leather.

Leather is strong and durable? Maybe so, but to scientists and technicians such terms are pitifully vague. You might want abrasion resistance in a shoe upper, suppleness in a bellows, tear strength in fine upholstery. Tour a testing lab like that maintained by Leather Industries of America in Cincinnati, crowded with wallets sliced open, belts torn apart, and jeans patches washed and rewashed, and you'll find instruments to test these and most every other property of leather. Attach a specimen to two pistons, forming a cylindrical bellows, set them rapidly oscillating for so many thousands of cycles, then inspect the leather for cracks. Expose a second specimen to a xenon lamp of controlled intensity, for a set time, and measure how much it fades. Clamp a third to a special cloth containing patches of wool, nylon, cotton, polyester, or other textiles, and see how much dye leaches out onto them. In other tests, described in standards established by the American Society for Testing and Materials, you measure the sample's chromium content, or spin abrasive wheels against it, or measure—ASTM spec D2214 applies—how well or poorly it conducts heat.

Here, where upholsterers and garment makers bring products to be tortured and tested, leather is not something to appreciatively run your hands over, but to get stretched, drenched, rubbed, pounded, and otherwise abused. And this has been leather's lot for most of its history. Leather has been used for gaskets, pump washers, clutches. For saddles, harnesses, and buggy whips. For the boots, belts, bags, and book covers for which it's still used today, but also, as in India, for storing oil and butter in bottles made of molded camel hide. In Britain, leather "bottels" were common for carrying water and wine. Leather was made into buckets to haul gunpowder, into coach springs, dog collars, and razor strops, into straps for subway riders to grasp as trains shuddered and lurched. Leather was used to sheathe swords, link the iron disks in chain mail, and cover shields—as in this translation from the *Iliad.*

Achilles threw in turn his ponderous spear
And struck the circle of Aeneas' shield
Near the first rim, where thinnest lay the brass
And thinnest too the o'erlaying hide.

Today, soldiers rely on body armor made of Kevlar, not shields of leather and brass. Electric motors have replaced flapping leather belts.

Books are bound in cloth much more than in leather. Bottles are made of pretty much anything *but* leather. Leather's range of use has shrunk, superseded by materials, superior by one measure or many, that have scant reason to imitate it; no one embosses Kevlar body armor to make it resemble leather. Today, where leather is esteemed the material of choice, it's not alone for its strength or durability but for the way it satisfies the eye, and answers to the touch.

In an early Bob Dylan song, a woman "sailin' away in the morning" asks her lover whether she should send him back "something fine . . . from the mountains of Madrid/Or from the coast of Barcelona." No, he says, he wants only her, wants her back. But in time he realizes her heart is roamin', that she's not coming back—and in the last line amends his reply: "Yes, there's something you can send back to me," he says: "Spanish boots of Spanish leather." *Lose girl, get boots?* Ah, but what boots! Going back to at least the 12th century, Spanish leather is tooled, highly ornamented goatskin, prized for its distinctive beauty; monographs have been written about it, a cologne named for it. Boots just to keep the poor fellow's feet warm? No, these were *beautiful* boots.

And that's how it is today with leather generally: In fashion advertising, websites, and car dealer showrooms we hear encomiums to leather—gushing hype, over-the-top fetishizing—that are never inspired, it's fair to say, by leather's thermal conductivity.

In a fluorescent-lit fifth floor room at the Fashion Institute of Technology in New York City, José Madera is helping a dozen students make slacks, blouses, and vests out of light garment leather. They lay out patterns, line up seams, cut notches, install zippers. But it's an introductory class, part of a certificate program in Leather Apparel, and some of them are struggling. The sudden high *twwinnng* of a broken needle silences a sewing machine's clatter. They lay out their collars upside down and wonder why one side of the seam is an inch short. Their seams ripple. They're students, and they're learning. And, like generations of graduates of this legendary garment industry training ground, their designs may someday adorn models on the runways of New York, Paris, or Milan. Mustard yellow leather pedal pushers? Tasseled leather chaps? Biker jackets? Not many are made in New York these days; the patterns mostly go to China for manufacture.

Madera, a veteran of twenty-two years in the business, just now answers a question, sits down at the machine himself to sew a troublesome seam. But this is his night job. Days, he's quality control manager for a company that makes leatherware in China, where leather sometimes goes for 75 cents a square foot. Fewer jobs for Americans, more for Chinese, Indians, Indonesians. But whoever's got the business, the suede vests, soft jackets, and clingy leather skirts coming over by the boatload offer sensual qualities people want. *Want for their bodies.*

Leather is a peculiarly androgynous material, melds male and female, can be as unyielding as a tree trunk, as welcoming as a flower petal. Harriet Beecher Stowe called the slave dealer in *Uncle Tom's Cabin* "a man of leather." Women who happily choose to remain unmarried sometimes call themselves "leather spinsters." In both cases the word implies toughness. But leather sheathes the lushest jewelry cases, is used for garments of the most sublime softness. "It's quite truly a second skin," says Michele Bryant, who also teaches at the Fashion Institute. Among potters, "leather hard" refers to unfired clay soft enough to carve or model, yet firm enough to handle. This, then, is leather's peculiar essence, its embodiment of contrary qualities—work boots *and* ballet slippers. Possessing both male and female energy, leather asserts its erotic charge.

And not just in short skirts and thigh-high boots, but in trenchcoats, chaps, belts, jackets, and other more workaday garments; no one hesitates to describe as sexy even a leather sofa, or a leather briefcase. Leather feels edgy and rebellious, says Bryant. And it's been that way since before Marlon Brando and James Dean draped leather jackets around their slim muscled bodies in the 1950s. Today, leather-garbed actors, rock divas, and models project sexuality in ways too familiar to mention. But so, even, does the leather industry itself. A trade journal ad for one of the big chemical suppliers, Rohm and Haas, shows a woman in a short leather skirt and leather boots, lots of leg showing, sprawled across a tufted leather couch. Another shows a doe-eyed miss, overalls concealing almost nothing of her bare breast, leather bag across her shoulder: "I'm in love with a wonderful hide—and it's Stahl finish, of course." Leather suggests raw sexuality kept barely in check, but also the tender, vulnerable flesh in all of us.

Leather's sexual energy, says Valerie Steele, draws from both its physical characteristics and its cultural associations. Steele graduated summa cum laude, Phi Beta Kappa, from Dartmouth, got her Ph.D. from Yale in European cultural and intellectual history. Today, along with being a Fashion Institute professor, she directs the Institute's museum, presiding over its tens of thousands of garments and quarter-million fabric swatches. She is also editor of a scholarly journal, *Fashion Theory*. What she thinks about for a living is fashion, clothes, and the human body. And among her interests is fetishes. She is the author of an Oxford University Press volume entitled *Fetish: Fashion, Sex and Power*. "By the beginning of the twentieth century," she says, "leather had acquired a fetish appeal." And for all the swinging of the fashion pendulum over ensuing decades, it kept it. It's "a constant as a fetish material, a fetish remaining unchanged for decades, centuries. It has incredibly strong associations with sex." And not just familiar white-bread sex, either.

It is the 22nd annual New England Fetish Fair and Fleamarket, a celebration of sadomasochistic sex, both homosexual and hetero, held on several floors of the historic Park Plaza Hotel in downtown Boston. The place is thick with booths given over to sex toys and bondage equipment; with half-clad men and women being led around on leashes; with panel discussions and demonstrations given over to topics like "Blood & Lust," "Fantasy Made Flesh," and "Florentine Flogging"; their presenters go by names like Boymeat and Mistress Sly. To some, this world of dark sexual practices bound up in surrender and submission can seem troubling, even perverse; it is not to everyone's erotic taste. Yet, a familiar air of crafts-fair-on-the-green runs through the Park Plaza, too, making it not so unlike other shows and conferences. Panels meet at set times, in gold-filigreed public spaces like the Clarendon Room and the Imperial Ballroom. Artisans and shop owners who look as if they'd be just as happy at a Renaissance Faire earnestly hawk their wares. And for every man wearing little more than a leather jockstrap led around by a woman brandishing a leather whip, there are twenty, unexceptionally dressed, who look like they've just piled in off a commuter train from the suburbs.

But all through the Park Plaza, one material, leather, predominates. Whips, chastity belts, collars, and harnesses? Clothing designed to expose the genitals or the breasts? Most are made of leather, typically black leather, leather all the time and everywhere. Salespeople tend booths with names like Leather Creations, or Larry & Leenie's Lusty Leather. Right beside brochures for sex toys, erotic jewelry, and bondage equipment, you find one for Tandy Leather Company, the Texas-based purveyor of family-friendly leather and saddlery gear, its "Nature Tand Basic Leathercraft Kit" offered for $29.99. Alternatives to leather, including PVC and latex, figure in this not-so-underground scene, too, but leather gives it its name. Look up LEATHER in one online encyclopedia, and you get familiar facts and definitions, along with an almost beatifically wholesome photo of leatherworking tools. But the entry refers also to leather as fetish object, and offers a link to LEATHER SUBCULTURE and its sadomasochistic sex practices. "Leather: It's not just cowhide," declares another website. "It's a lifestyle."

Why leather's hold on this particular corner of the erotic imagination? When I ask around at the Fetish Fair I encounter no want of theories. Leather is *skin*, and so innately sensual; all agree on this much, at least. I hear how thick leather jackets, boots, and other protective riding gear became ubiquitous among post-World War II motorcyclists, whose breed of sometimes exaggerated masculinity appealed to many in the rough sex crowd. How "Tom of Finland," born Touko Laaksonen in 1920, a Finnish illustrator who devoted his creative life to depicting patently aroused, tight-buttocked men in black leather boots and motorcycle jackets, left an indelible erotic imprint on this culture. How the dominant figure in ritualized bondage sex play, the "top," can be seen as ruler or prince, for whom the luxury of leather and the intimidation of black is his due. How the "bottom," his submissive partner, can be seen as slave, or beast, rightly trussed up by just the sort of thick leather straps and harnesses used to restrain animals. In trying to explain its role as a fetish object, Valerie Steele has described leather's erotic spell as "overdetermined." By this she means that its feel, sound, and smell, its link to "animalistic and predatory impulses," all play a role—but not in any simple way; its sexual charge bubbles up from the culture around it, and from leather's own nature, in ways impossible to tease apart.

But if leather exerts so primal a hold in this sexual netherworld, should we be surprised that it carries an erotic charge, if a tamer and more PG-rated one, among the rest of us? Certainly it is not just fetishists who love it, or value it for its sensuality.

Consider, for example, Levenger, the Florida-based mail-order company that sells "tools for serious readers," everything from pens and notebooks to bookshelves, chairs, and lamps, often with a retro flavor. In its catalog, much as at the Park Plaza, leather appears on almost every page, no small effort of the copywriter's art devoted to its sensual pleasures.

Levenger wallets and document cases, we learn, are available in belting leather, which "wears like a good pair of leather chaps, from its greenhorn tone when new to the burnished glow it achieves with use." Its Palm Pilot covers come in "a smooth, glazed leather that indulges your hand." And do you know those manila envelopes you seal by winding a red cord around a little cardboard button? The kind that set you back maybe 89 cents at your local stationer's? Well, Levenger sells one in red leather that "will burnish beautifully the more it's handled," and sells for a 100 bucks.

Or consider the woman's handbag. "There isn't a human alive who doesn't get total sensual overload upon opening up a brand-new luxury leather handbag," Valerie Steele and coauthor Laird Borelli quote one enthusiast in their book, *Handbags: A Lexicon of Style.* Among the finest handbags, of course, a great many are sheathed in leather, or lined in it, or both. Hermès started out as saddlers to the aristocracy, Fendi as a purveyor of luxury leather goods and furs. One handbag fashionista, Claire Wilcox, refers to "the curiously personal and intimate nature of the handbag," with its distinct interior and exterior. Another writer who has made the subject her own, Anna Johnson, writing in *Handbags: The Power of the Purse,* reports that "psychiatrists have always held the bag in suspicion, imagining it to be a 'vagina dentata'—the only place a man's hands are unwelcome." Intimacy, then, and luxury—which to Steele and Borelli imply indulgent, voluptuous pleasure: What better terrain for the softest, most exotic leathers, the most select skins, the material itself contributing much to the handbag's allure.

For purveyors of wallets, handbags, shoes, and garments, it's never long before it's the leather going into them, *the leather itself,* that's being lauded, its silky touch or rich sheen made to reflect back onto the product. Shoes: "Cobbled in Italy, with lustrous, hand-finished Italian calfskin uppers, full leather linings, leather-covered insoles, sueded leather soles, and leather/rubber heels." Desk binders: "Made of top-grain cowhide that's drum-milled to a plush, pebble-grain texture and pigment-dyed for rich, long-lasting color." In catalog copy, in sales patter at boutiques, shoe stores, and crafts shops, it's lovers of leather, as a material, to whom they appeal; I'm one, and maybe you are, too. One company openly acknowledges this self-aware consumer niche. Its catalog points to a leather-cushioned dining room side chair from a top designer: "Go public with your affinity for leather."

Leather exerts its hold on not alone those who buy and sell it, but those who work it. "'Feel this,'" Philip Roth's protagonist in *American Pastoral,* known as the Swede, son of a New Jersey glove manufacturer, recalls his father telling him as a child,

> and the child would crease a delicate kidskin as he'd seen his father do, finger the fineness appreciatively, the velvet texture of the skin's close, tight grain. "That's leather," his father told him, "Look at the pores of this skin with a magnifying glass and they're so fine you can't even see 'em."

Roth describes with equally affectionate detail how glovers worked with kidskin. "The cutter has to visualize how the skin is going to realize itself into the maximum number of gloves," the Swede tells a visitor. "Then he has to cut it. Takes great skill to cut a glove right. Table cutting is an art. No two skins are alike. The skins all come in different according to each animal's diet and age, every one different as far as stretchability goes. . . ."

That was gloves, and in Roth's magnificent telling, there is love there, in the father, in the son, in their world. One's attraction to a craft rests largely on one's affection for some particular material, writes Hart Massey, a contributor to *The Craftsman's Way,* a compendium of Canadian crafts lore. "The potter is turned on by clay, the glassblower is excited by the 'power and magic' of molten glass, the weaver loves wool, the blacksmith has an affinity for hot metal, the jeweller is fascinated with precious metals, the shoemaker enjoys the sensual qualities of leather." Another contributor, David Trotter, a designer and craftsman from Beeton, Ontario, sees leather as a notably "responsive" material. "Any kind of mark that you make on it, any way that you shape it when it's wet, it will respond immediately. Yet it will retain what you have done indefinitely." Writes a Saskatchewan shoemaker: "I feel I could be lost in leather for the rest of my life and not really scratch the surface of it."

The men and women who make saddles and boots, wallets, vests, and gloves, whether artisans in small shops or workers in large factories, cut and stitch leather. They skive it, or thin it down. They pierce and punch and tool and stamp it. They rivet it. They lace and braid it. They laminate pieces. They mold leather. They glue it. They dye it. The saddler builds his work on a wooden "tree." In early 17th-century Spain, they used wooden molds and counter-molds to emboss leather panels with intricate patterns. Workers in a Coach factory in New York, featured in a company-published book before their jobs went offshore, used industrial grade sewing machines; and heavy presses that bore down on steel dies to punch out individual pieces; and open flames, that burned off stray threads left in sewn leather belts.

But, as the title of one frankly adoring photo essay on the making of a Hermès handbag put it, "In the Beginning, There was Leather . . .". *The leather itself.* The first photo in this *New York Times Magazine* article shows the skins, piled on the floor or on wheeled dollies, sorted by color. "In the workshops of Pantin, outside Paris," we learn, "the skins await inspection." Only those earning Hermès's gold seal, whatever that is, end up in one of its $5,500 bags. More for leather, perhaps, than other materials, it's often like this: Whether in a custom shoemaker's atelier in Budapest, or a Coach factory, or any of a thousand bustling little shops in China or India, in the beginning there is the leather, with its smell, its feel, its peculiar sensual seasoning.

In 1972, a group of leather craftsmen produced a book, *Leather*, edited by two of them, Donald J. Willcox and James Scott Manning, that introduced the craft, offered advice on how to get started in it—and, along the way, served up a distinctly Sixties-flavored manifesto. "This book," they declared, "is a trip—a leather trip." People were using their hands again, they said, creating a new crafts movement, and leather was at the spiritual and moral center of it; forget about the dreary leather "kit-ism" of summer camp and school crafts fairs. The book portrayed hippie girls in tight-fitting trousers with laced seams, full-bearded men in leather hats and fringed suede coats. It showed the dark, wood-planked interiors of leather studios, on-the-fly leather shops set up under Indian print awnings at the ocean's edge, leather formed into outlandish masks, decorated jewel boxes, dolls, and dragon-figured wastebaskets.

For the authors, full of that youthful Sixties freshness and wonder that can seem today merely quaint, leather wasn't just a craft, a way to make a living, but a movement and a quest. People were drifting

> into leather. All kinds of people. Serious people. Some of them were looking for an alternative life style through their hands; others had been participants in change, but had been unlucky; and still others were ordinary, quiet, hard-working people. These people had many different backgrounds, but the vibrations into leather were pretty much the same.

> And when they began—ZAP! All kinds of things started to happen. Sandal and leather shops begin to spring up all across America. . . . After the first fires were lit, the flames spread quickly.

It was during this era, and under the influence of just such sensibilities, that I began working with leather myself. A few years earlier, when I was in my early twenties, I'd had a suede sportjacket made for me. From a bar waitress wearing a rustic leather vest, I'd gotten the name of the woman who'd made it for her and arranged to meet her at a downtown Baltimore department store to select a skin. Peggy measured me and when, a few weeks later, it was nearly ready, I went out to her tiny stone cottage on a dirt road outside Baltimore for the final fitting. It was a beautiful deep brown suede, so soft you sank into it, and when she draped it in all its fullness around me and I felt her strong young hands on my shoulders and back, the little charged thrill of it insinuated leather into my life forever.

Then, a few years later, in 1973, living in San Francisco, I needed some sort of bag or briefcase to carry my books and papers. I'd admired some of the rugged leather ones men were beginning to feel comfortable carrying, but couldn't afford one, and so resolved to make one myself. I walked into a dark, scent-filled leather shop off Columbus Avenue in North Beach, talked to the proprietor about what I wanted to do, and soon was being schooled in the basics. I walked out with some thick harness leather, a rotary punch, a skiving tool, a sewing awl, and a few other tools. It was no work of high craftsmanship I produced but in the end I had a serviceable and—may I say it?—handsome over-the-shoulder bag with brass fittings. It served me a long time. Only after a quarter-century of almost daily use, with the leather finally tattered beyond repair and even the brass hardware worn through, did I retire it, a few years before starting work on this book.

I also made a few women's handbags, and a funny double-lobed key holder for a friend who was always losing her keys. And a notebook cover out of pebbly black leather. And a small briefcase from a waxy orange-brown hide that, I realized only later, showed every mark. And saddlebags for my bicycle designed to look more at home in the office than on the bike and that still gets compliments from fellow cyclists.

Once I made a pair of shoes—regular street shoes with thick, two-piece leather soles, hand-stitched, made without pattern or last, modeled from my own feet. It was trial and error all the way. I'd sit there with tools on my lap, rough-cut pieces of leather draped over my foot, lengths of waxed thread dangling from unfinished seams, innocent of any idea of what to do next; I only had two hands and could have used three or four. The triumphant moment came when I completed the soles. Up till then, the uppers had as yet been sewn only to the topsole. But now I glued topsole to bottom, pounded them together with a wooden mallet, and drove cobbling nails through three layers of leather into a thick steel backing plate that bent them around to form a clinch. Then I trimmed the edges, sanded them smooth, and polished them. All that remained was to initiate my handiwork to city streets and glass-strewn sidewalks. I skipped down the stairs of my building, All Feet, no longer aware of the rest of me, my new shoes shooting off sparks of pleasure at being shoes, dancing to the beat of the afternoon traffic.

I never became a consummate craftsman; I can only imagine what a real pro would say of my clunky efforts through the years. Still, the work gave me pleasure, and so did the material. I'd visit leather shops, run my hands over the hides, fold them over, let them flop back. Thin and supple or thick and crusty, it didn't matter; I liked them all. Today, I haven't the years-bought knowledge to select the perfect hide for a project, or skills refined enough to superbly execute it. But in my long friendship with leather, I find it forgiving, strong, and deep. The things I made I mostly still have, and use, years later, and they still give me satisfaction.

Pleasure in the accomplishment? Yes. But I enjoy, too, the sheer beauty, to eye and hand, of the material itself. It wears well. It never fails precipitously, like polyurethane that delaminates, or vinyl that, tearing, reveals the coarse scraggly cloth beneath. But only gradually, the leather sueding up where it's worn; or else growing pipy with use; or drying out. You can polish over the worn spots, though, and oil it up, replenish its moisture. It doesn't look new, mind you, but gracefully old. Like a beautiful older woman, skin a bit wrinkled, eyes aglow with life.

Up on the shoe factory's fourth floor lie the graded hides, mostly calf, gathered in neat piles, or rolled up and mounted on wooden racks. In the adjacent cutting room stand racks loaded with dozens of dies, each a steel cookie cutter shaped like the piece to be cut out, made from the original patterns. A workman, the "clicker," as he is known, selects those he needs.

He lays out a hide on the bed of the old machine, positions dies atop it, and presses two buttons (both hands thus removed from harm's way); the heavy arm suspended above swings down onto the die, punching out the leather, swings back up. Then, in a blur of motion, he moves dies into position over another stretch of leather, for another cut. This is fast, skilled, high-paid work. He fits the long axis of each piece of shoe upper perpendicular to the backbone, to reduce unwelcome stretch. Makes sure the leather of the left shoe looks like that of the right. All the while, tries to make best use of the hide, use as much of it as he can, yet avoid brands, veins, tick bites, and stretch marks—including any barely visible ones that might become ugly and obvious once pulled around the last. Each cut is a decision, a balancing act of yield vs. quality. Avoid every last blemish or defect? No, too much waste. Use every last piece of hide, heedless of surface defects? That won't do, either: You'll get production problems, along with blemished shoes customers won't buy. Finding a middle ground is the cutter's art.

This old, creaky-floored factory in Brockton, Massachusetts, makes men's shoes. And making shoes means that every smallest feature of the final shoe, every curve, every stitch, every perforation, hole, dimple, or design, every slather of glue, every lining, interlining, eyelet, or seal, has to be *done*, is the result of some worker's operation, hundreds of them. In a custom shoemaker's shop, they would mostly be done by hand. But here they're done easier and faster with special-purpose machines. Small, rapidly spinning wheels bevel and thin surfaces where they must be stitched or cemented together—down from the thickness of six playing cards, say, to that of two; otherwise you'd wind up with thick, ungainly clumps of material. Quick-fingered seamstresses spin leathery clumps in hairpin curves around frantically pulsing needles. This is no assembly line operation. There are virtually no conveyor belts here. It's a batch process, the batch being a dozen pairs, known as a "case"; a pink tag accompanies each through the factory, recording the sizes and widths of each shoe that makes it up.

For most of the manufacturing process, what will become a shoe resembles nothing so much as a shapeless mess of leather and stitching, aimless threads and glunks of glue, floppy scraps. It looks like a shoe only once it comes off the last; the word goes back at least a thousand years to the Old English *laest*, which meant foot shape or footprint, but the object itself is known to have been used by the Greeks and Romans. Traditionally made of wood, now often of heavy plastic, the last is the object around which the shoe is constructed; it gives shape to the shoe. For eight hundred dollar custom-made shoes, the shoemaker measures every dimension of the foot—across the instep, heel to big toe, and so on—and carves the last from them. For mass production, each size and style gets its last, which is thus an approximation to your foot or mine.

The upper, sewn together now but shapeless, is first spritzed with heat and mist, as if you were steaming a stale roll to freshen it up. Then it's inserted into the lasting machine, progenitor of one patented in 1882 by Jan Matzeliger, the brilliant son of a white engineer and a black slave; when introduced to the shoe-making center of Lynn, Massachusetts, it was reputed to have halved the price of shoes, increasing their output by a factor of twelve. With the last now in place within the bowels of the machine, pincers grip the leather from beneath, and pull it down around the last, onto which the whole assembly is tacked into place. During this iconic moment in the making of the shoe, the leather itself, that essence-of-shoe, seems lost among sliding steel parts and pneumatic and electric lines. Once, before becoming leather, the hide had covered a living, three-dimensional body. Then, from the tannery, it came out flat, two-dimensional. Now, from the lasting machine, again three-dimensional, it's taken on smooth, unpuckered contours that will cover the foot of another living body. It's a violent process, the leather being stretched, tucked, and tortured into shape. Once it's over, better believe it, the leather has proved its worth. Leather must be strong for the sake of the foot it envelopes and protects? Yes, but it must be strong, first of all, to survive being made into a shoe.

Back in 1912, America led the world in export of leather shoes, holding 55 percent of world trade. In the early 1960s, before Corfam, when China was Communist in more than name and before it dominated the world's footwear manufacturing, factories in New York, New England, Pennsylvania, and elsewhere in the United States turned out six hundred million pairs of shoes a year, three-quarters of them with leather uppers. Few American shoe factories remain today; by early in the 21st century, annual production of shoes with leather uppers in the United States stood at barely twenty million pairs, about a hundred thousand of them from this one in Brockton.

Manufacturers have long made shoe parts out of nonleather materials. A 1916 account tells of celluloid and oilcloth used for toe boxes, and "leatherboard"—fibers of hard leather, waste paper, rags and wood pulp, rolled into hard sheets—for soles. In the 1940s, when the federal government brought an antitrust suit against the United Shoe Machinery Corporation, it found the company exploring a variety of leather substitutes—from porous plastic insoles to plastic heels, plastic eyelets, plastic welts. Beginning especially in 1944, Goodyear's Neolite and similar materials, with names like Panoline and Maxecon, seriously challenged leather for soles, by the late 1960s going into nine in ten of them. And then of course there were sneakers, the name coined in 1917 to help sell the mass-marketed Keds; its rubber soles presumably made them stealthy. These progenitors of today's athletic shoes, which used lots of rubber and canvas but no leather, became a big business. Children and teens, especially, went for them—enough so that by 1961 Leather Industries of America was taking out ads that urged parents to "insist on shoes with leather uppers and leather soles" during their kids' formative years. Heels, puffs, adhesives, as well as soles were, during this period, all being made of plastic; for these shoe parts, polystyrene, polypropylene, and polyethylene were better than leather.

But men's and women's *uppers*?

Neolite soles succeeded for functional, not aesthetic reasons; that moment in the shoe store when you relish the sight of a new, oak-tanned leather sole, before it's unrecognizably abraded by use, may be the last time you ever look at it. Not so for uppers, the star of the shoe, the part expressing its style that you see each time you put them on and that others might admire as well. If you were a man and wore a brogue, if you were a woman and coveted a classic Courrèges square-toed Mary Jane or a Perugia pump, it was upper leather that embodied these styles. Uppers consumed one to two square feet of calf or cowhide per pair, and used more than half the leather made worldwide.

Substitute materials like vinyl made barely a dent in the market for women's and men's dress shoes. Despite other, more successful assaults on leather, here tanners could cling to an air of superiority; leather occupied its castle, secure within its walls. Shoe uppers were leather because they'd always been leather. Calf, cowhide, deerskin, alligator, but most always leather. By nature, by the weight of tradition, by enshrined industrial practice, leather was the best material for shoes, everybody knew it, and that was that.

Pressed, leather's champions would point out that it adapted better than any other material to one critical, if not precisely ennobling, fact of human physiology: *Feet sweat.* It is no revelation to assert this fact, of course. But the author of an 1847 book entitled *The Book of the Feet* cited a Dr. Erasmus Wilson to just this effect. It seems that Dr. Wilson had counted the number of "perspiratory pores" per square inch in the palm of the hand, and on the heel of the foot. The figure he arrived at, some thousands per square inch, was close to that established by scientists most of a century later. This vision of perspiration in full flood apparently inspired in Wilson a nightmarish second thought: "What if this *drainage* were obstructed?"

Feet sweat, the moisture they produce *can't* be obstructed, needs to be spirited away. Leather does that brilliantly; it *breathes.* This was the incontrovertible fact that tanners, leather chemists, and shoe industry men invoked as leather's badge of superiority during the 1950s and early 1960s, even as Du Pont engineers worked on their synthetic microporous material, the future Corfam. *Three hundred square meters in each gram of shoe leather!* That's how much surface area those bundles of collagen fibers offered to the transmission of water vapor, a National Bureau of Standards scientist reported in 1950. Yes, leather shoes protected the foot; wet or dry, leather offered better resistance to nail puncture than any other material in use, claimed an article in *The Leather Manufacturer* in the early 1950s. But it was all that surface area, in all those collagen fibers, that made leather unique; it was what "substitute materials do not possess." The authors' confidence and certitude warms their otherwise data-stuffed technical account. They didn't say it but could just as well have: *There is nothing like leather.*

> A few pits lay like a lime-splashed chessboard in the yard where water, tanning liquor, the greenish depilatory liquid and blood trickled between the stones. Men would be working the gulping pumps, leaning over the fleshing boards, standing before the screaming splitting machines, and the tumbling drums, wheeling the green pelts which quivered like jellies into

> the sheds, standing by the stench of tubs packed with fermenting dung, while above their heads the belting loosely wangled round and round and shook the floors.

This scene from an early V. S. Pritchett novel depicts the tannery that is its main setting. The acclaimed British novelist and short story writer died only in his nineties, and was remarkably prolific, but got off to a late start; he'd spent his earliest years as a tannery apprentice in London. And this, his second novel, published in 1935, is imbued with his experience. It bears the title *Nothing Like Leather*.

The phrase, "There's nothing like leather," today carries no taint of irony. It—and variants like "Nothing Takes the Place of Leather," the title of a 1924 booklet by the American Sole & Belting Tanners—appear in virtually every encomium to the material by tanners. John Waterer, one of leather's most dedicated champions in mid-20th-century Britain, affirms the old adage in his 1946 book, *Leather in Life, Art and Industry*:

> There *is* nothing at all like leather—there never was, and, it is fair safe to say, there never will be. Man is unlikely to attain to the *ars perfecta* of Nature; and whilst it is true that leather is a manufactured product, impossible without the art of man, its basis is a product of Nature of wondrous structure and beauty.

For many old leather hands, then, the adage simply asserts a self-evident truth, that to imitate, replicate, synthesize, or substitute for leather is in the end fruitless, the discussion over.

But the phrase did not always mean what it means today. It apparently first appeared in Daniel Fenning's *Universal Spelling Book*, an English primer dating to 1767 that ultimately went through thirty editions: A town, it seems, is besieged. The town council meets to discuss what to do. Make walls of good strong stone, says a mason. Use wood, says the carpenter. No, a currier gets up to say, "There is nothing like leather." Here, the proverb does not celebrate leather at all, but comments wryly on the human penchant to stick to mental ruts—to think, we might say today, very much *inside* the box.

Pritchett's novel is about a not entirely sympathetic working class character, Mathew Burkle, who all his life covets the tannery he runs but cannot own. He is a limited man who, overwhelmed by aspects of himself and others he doesn't understand, retreats to work and its certainties. "Go away and work," he says to himself at one vexing moment. "Make leather. There is nothing like leather." In Pritchett's telling, this says nothing of the material. Rather, delivered mechanically, it acknowledges that part of Burkle clings to the safe, the predictable, and the workaday.

We saw in Chapter 1 how, at the time of Corfam's introduction, Leather Industries of America warned consumers of the danger of shodding their feet in anything but leather, one ad turning once more to the trusted, time-worn maxim: THERE'S NOTHING LIKE LEATHER. In doing so, they meant to praise it. But ironically, their claim embodied all the sluggishness of an old, hide-bound trade challenged by new ways and fresh thinking.

7

"All Shortcomings Have Been Eliminated"

i. "The Interesting Work of John Piccard"

"I have been acclaimed as the inventor of Corfam," he says. "I am not."

At the age of 83, John Piccard speaks Truth as a train of declarations and assertions. He looks like a Victorian patriarch, with a long white beard and a full head of straight white hair. He speaks in complete sentences and paragraphs, forswearing contractions. He has a story to tell and he tells it.

It is not hard to guess that he might have been an outsized presence at Du Pont when he was 32, which is about how old he was when he did, or did not, invent Corfam. What did he look like back then? "You'd better ask what he *sounded* like," replies Hamilton Fish, an engineer who worked with Piccard at the Experimental Station in the 1950s. What he sounded like was booming, forceful, and sure. "He always talked 4f," says Fish, referring to the musical notation, beyond even *fortississimo*, that means extremely loud. "He would walk along the hall with two or three others and you'd hear him sounding off like a locomotive."

But oh, the way his ideas billowed forth. "He was a tinkerer, odd and creative," says Fish, though he sometimes balked at shepherding ideas too far along the gray path of practicality. "'That's not what I do,'" Fish has Piccard saying. "'I invent.'" "He was the last Renaissance guy," says Joe Rivers, his boss for a while. "He knew about everything. He knew mechanical engineering, chemistry, aeronautics. He had an interest in any technology that came his way. And he was always coming up with new ideas." Rivers had gotten his Ph.D. in chemistry at MIT in 1942, then gone straight to Du Pont. When Johnny Piccard joined him in Buffalo, New York, in 1949, Rivers was heading up a research group busy dreaming up uses for polyester and other synthetic fibers. There were maybe half a dozen of them, with a range of expertise in fiber-spinning technologies. Piccard reported to him, says Rivers, "on a very freewheeling basis—as much," that is, "as he ever reported to anyone."

Piccard was born in Canton Vaux outside of Lausanne, Switzerland, on July 25, 1920. His mother was from old Yankee stock, his father Swiss, his grandfather professor of organic chemistry at the University of Basel. One day when he was small, his grandfather took him around to a machine shop, all lathes, the smell of burning oil, and the looping leather belts that drove the machines. *They're making things here*, Piccard remembers thinking. He went home and told his mother he wanted to do that when he grew up. His mother thought, *Oh, how nice, an engineer.* No, he meant, he wanted to be a machinist.

He came over with his parents when he was 6. He remembers wearing, at age 7 or 8, a black Fabrikoid coat with a real sheepskin lining and collar. Fabrikoid, his mother said, was better than split leather, not as good as full grain; that's just what Du Pont said, too. Later, Piccard enrolled in the University of Minnesota, where his father was a professor. But with the war he trained as an ordnance officer, he reports, and did antisubmarine work with the Eighth Air Force in England.

When he finally graduated in 1947, he says, he impressed the Du Pont recruiter with his account of a studio-quality recording lathe he'd built and about how he'd worked up a fluorescent ink that, stamped on your skin once you'd paid to get into a dance, unmistakably identified you for the evening as PAID. "They offered me a job," he says, "and I took it." The job was at Du Pont's Pioneering Research Laboratory in Buffalo—so named because it was dedicated to advanced research—presided over by the inimitable Hale Church.

Ten years before, Church had figured out how to keep things fresh with cellophane wrap. His innovation soon sheathed most of the cigarette packs of America, was a booming commercial success, and got him a big lab where he could do pretty much as he pleased. Church relegated dreary financial matters to a lieutenant, remembers Rivers, freeing him to "get out on the front line and talk to you for hours" on matters chemical, sparking your imagination. And he fairly embraced the chanciness inherent in research. "He didn't give a damn if we had a failure. He'd rather have a success, but he was very happy to let people follow things that looked highly unpromising." In 1949, Charch wrote to a Du Pont colleague that they'd need to "dig up all the ideas [they] have and put them down on paper"; he was talking about plastic films. "For the present at least, no brakes, ceilings, or fences will be applied to the work."

When Piccard stepped into Charch's research fiefdom in July 1947, he was put to work on Polymer V. Polymer V was polyester—Dacron before there was a Dacron, the result of a tortuously contested technological history which left it by 1947 a centerpiece of research at Pioneering. "It was the hottest thing in the laboratory when I came along," says Piccard. Everyone was working on it. Synthetic paper was one possible use for it. Another was nonwoven fabrics.

Think fabric. Think of a loom, its shuttles carrying threads back and forth across the fabric's width, making for the age-old warp and woof of common textiles. Or think of the clickety-clack of knitting needles, making sweaters and scarves. But there's another kind of fabric, one as familiar as the felt on an old hat band: not woven, not knit, but built up from a more or less random array of fibers bound or fused together. Since as early as 1944, Charch's lab had been working on such nonwoven fabrics, and it was on nonwoven fabrics that, at least by November 1948 and probably earlier, John Piccard found himself at work. "Nonwovens," he recalls, "were everywhere, much heard about in watercooler conversations," countless ideas floating about for how you'd make them. "Blue sky loose talk," Piccard calls it.

He began tracking down samples of any fabrics he could get his hands on—unbleached muslin, vinyl sheeting, waterproof tarpaulin, oak-tanned sheepskin, canvas, burlap, paper sacking; it would be good, he decided, "to know something about fabrics in general." He'd weigh them and test them for tear strength, abrasion resistance, burst strength. Then he started playing; experimenting was what it was, if of a primitive sort. In his early nonwovens work, he didn't start out to make a leatherlike material, certainly not a leather that breathed. Rather, as he wrote in a progress report in August 1949, he was after "any possible novel and interesting constructions or combinations of materials." *No brakes, no ceilings, no fences.* Materials "with strength, softness, porosity, and durability." Some turned out clothlike. Others, especially when a little thicker, were more like leather.

"Making the initial leather, spreading the fibers out, doing something to get them to stick together, yes this was like playing," he says today. He used rayon, nylon, polyester. Fibers one day 8 inches long, the next maybe a thirty-second of an inch. Change the binder or how much of it. Variously mix. Set the samples out in a press, its platens heated up to 150 degrees Celsius, like a warm oven. See what you get. Test it. The result sometimes strong but not durable. Sometimes scraggly. Sometimes like paper. Sometimes more or less leathery.

"Probably very soon after we started, we recognized a similarity to leather." Could they get something like its porosity? In one early success, he made a mat of granulated polyethylene, polyester fibers, and cellulose acetate fibers, and fused them together in his press. Then he washed out the cellulose acetate, which was soluble in acetone. That took a while; the soluble acetate fibers, bound up as they were in a kind of matrix, offered scant access to the solvent. But slowly dissolved out, they mutated into wormholes, making for a leatherlike material whose porosity seemed to contribute to its softness.

Leather, or fabric for that matter, wasn't desirable just because it was strong or resisted tears. Mostly, people liked it soft. What exactly *was* soft? "It depended on what corner of the ceiling you were looking at," says Piccard. But he was an engineer, needed numbers, and so devised his own softness tester to supply them. You make a pocket of the test material, suspend it over a hole or ring of set diameter, and start dropping lead pellets into it. More rigid materials needed more lead to force them through the ring, softer materials less. The weight necessary established what he called its "ring softness value," which he claimed was reproducible and "a good measure of the 'hand' softness of the material."

He wasn't finished. Leather wasn't entirely homogeneous. It had a surface grain and beneath it a fibrous tangle; Piccard made a first attempt to ape that structure. Its surface layer was just soluble fiber in a polyethylene binder, in approximately even proportions, with no strengthening fiber. The recipe for a 9-inch square of its bottom layer, was 7 grams of polyethylene binder, 11 of polyester, and 22 of soluble acetate fiber. Lightly abrade the inner face, Piccard found, and it suggested suede. Dye the polyester, pigment the polyethylene, and you had a leathery brown. His sample, Piccard reported, was "soft, springy, and tough. It is particularly interesting because of its porosity, high absorbency, and good 'hand.'" Porosity, of course, meant it could transmit water vapor. Embossing a leather grain before washing out the soluble fiber, he asserted (though maybe this was just a guess), would probably little affect its other properties.

Now, this was primitive stuff, distant indeed from what would become Corfam; it's not even clear Piccard ever actually measured its breathability. Still, he'd fashioned something porous by virtue of its wormholes, pliable due to its plastic binder, durable due to its coating, strong thanks to its matted polyester fibers. So when, during Corfam's commercial honeymoon 18 years later, Du Pont was handing out recognition and handsome cash bonuses, Piccard was remembered: He had "observed and aggressively demonstrated that these non-woven structures . . . [had great] potential value as leather replacement materials" that could, presumably, be "fabricated into shoes with comfort features in the range of leather." It was a start, and everybody, even then, could plainly see it. The Corfam project would take many a turn between 1949 and its launch in 1963. But Piccard had set it in motion.

Set it in motion? *Fired it up* is more like it. As early as mid-1949, Hale Charch had his sights fixed on leather and was writing Du Pont headquarters in Wilmington for "certain information relating to the economic position of leather," such as the various types of leather, how much they cost, how they were used, which of them might prove plentiful or in short supply. He suspected they were onto a commercial goldmine but couldn't be sure.

> You have seen laboratory-prepared samples of leather-like fabrics produced from Pioneering's work to date. Without having any specific knowledge of the leather business we have, temporarily at least, decided to point this work in the direction of fibrous leather substitutes on the *assumption* that there are relatively high priced markets here for a product that does not yet exist, which has the softness, the porosity or breathability, and something approaching the durability of, leather. In our request to you, we are really trying to determine [whether] this assumption has a basis in fact.

In August, Charch brought in a tanner named Weldon whom he and two other Du Pont men grilled about leather. Much of this was basic stuff—that the thickness of leather was measured in "ounces," equivalent to a sixty-fourth of an inch; that a side of leather was half a whole hide; that a split referred to slicing the original hide into two or more thicknesses; that a pound of raw cattle hide might yield a square foot of finished leather; that you lost about 30 percent of the hide when you cut it up into shoes. Pioneering had started out virtually ignorant about leather, but Piccard's work had given the division every reason to learn more.

In December 1949, Ray Houtz, who eight years before had helped develop what would ultimately become Orlon acrylic fiber, wrote to Church summarizing his thoughts on Pioneering's research program. Artificial leathers and other nonwovens, he said, were near the top of his priority list.

> This field, which has been reopened by the interesting work of John Piccard, has, in my opinion, very great potential. . . . I can visualize synthetic leather entering and taking over the leather market to a very great extent.

There was urgency to things now. As Joe Rivers remembers it, Piccard's "beautiful poromeric materials" (the word poromeric, Du Pont's coinage for its porous polymer, would come only later) were far enough along "that an [emissary] from Fabrics and Finishes came to talk to us about them"; F&F was the longtime home of Fabrikoid, the company's first artificial leather. Houtz was worried about potential competitors and deemed it "important that we get into this promising field on a larger scale of research effort as quickly as possible."

In mid-May of 1950, Piccard asked to scale up his work, requesting laboratory versions of production line equipment for forming nonwoven webs.

In June, Pioneering moved en masse from Buffalo to Wilmington; Piccard calls himself "the first technical occupant of the 302 building," a big brick laboratory just opened near the middle of the Experimental Station campus.

Almost before he got there, he was off with Joe Rivers to Newburgh, to talk to F&F about its work with coated fabrics. There they found interest in some of his nonwovens, for their potential as high-tear-strength improvements on pyroxilin and vinyl for coating fabrics, but little in the porous ones; maybe some day, Newburgh's Dr. D. E. Edgar allowed, but not now. Porosity was overrated, he as much as said. Why, plenty of shoe leather wasn't very porous at all. Another Newburgh man pointed out that comfort depended on moisture absorption as well as moisture transmission.

Back at the Experimental Station, Piccard's success at creating pores in plastic sheet material had them imagining a degree of porosity that might mimic that of leather itself and equal it for foot comfort. This subject became the preoccupation of Boynton Graham, a 39-year-old Ph.D. organic chemist from Haverhill, Massachusetts, in the heart of New England shoe and leather country, who began to make himself into an expert on it. In June he met with Piccard and another Du Pont researcher to discuss Piccard's "leatherlike compositions." The porosity achieved thus far was actually, uh, a little much. The pore-forming material, the cellulose acetate, consisted of fibers about as thick as those of ordinary soft cotton. And once washed out, it gave you pores big enough to let water through. Not, mind you, water *vapor*, in the gaseous state, which was of course the idea, but water itself. Cold, drenching droplets. One idea was to use much finer fibers that, washed out, would leave the material porous enough to let water vapor out but not enough to let water in. They had a ways to go.

But it was progress enough that by late September 1950 Dr. J. E. Evans, Piccard's colleague at Pioneering, was ready to put on a little in-house slide show, to acquaint others at the Experimental Station with progress on nonwovens and, as he said in his notes for the talk, their "very unusual properties." He described what Piccard had been up to, showed how he'd worked up samples with his little hot-plate hydraulic press. "It seems to lend itself to mechanization," he added tellingly, going on to cite the garnetting machines it would require, the calendering rolls, the embossing equipment. Piccard's "playing" of the previous two years already had people thinking about full-scale production.

Evans divided the work done so far into impervious nonwovens, the sort Newburgh was interested in, and "the leather-like porous type. These," he explained, "are distinguished by high internal porosity, softness of handle even in heavy thicknesses, and permeability to moisture diffusion. Comfortwise, these materials are equivalent to suede" or, rather, would be if they could get it to work. You'd use polyethylene as binder, combine it with tough polyester fibers blended with the soluble pore-forming fiber. The heated calender rolls would fuse them together. Embossing rollers would impart a leathery grain. Finally, you'd leach out the soluble fiber. *Voila*, a porous nonwoven, in all its leathery glory.

Finally, Evans looked to commercial prospects. The low-cost non-woven might go into tents and awnings, while "the deluxe porous or leathery type is directed toward the market for suede, pigskin, cowhide, upholstery coverings, fashion accessories, and some military usages."

He didn't mention shoe uppers, but shoe uppers were plainly on Du Pont's corporate mind. On November 15, Boynton Graham headed up to Philadelphia to meet Michael Fiorentino, a maker of custom shoes with a shop on Sansom Street near Rittenhouse Square; custom shoes were just what Graham wanted from him. At a time when good men's shoes went for $15, Graham was paying Fiorentino $45 a pair, for a plain capped-toe style, in $10^1/_2$D, similar to a popular Florsheim model. He supplied Fiorentino with perforated nylon-coated fabric as well as similar fabric without perforations, "designed to serve as a continuous and presumably uncomfortable control material"; it had already been tested as a shoe material, arousing complaints of hot-weather discomfort. But it was the four porous fiber mats, of varying composition, made by Piccard, each brown and 11 inches square, that were the real interest. Fiorentino was making the first shoes to use anything like the material that would become Corfam.

Not long before, Graham had scouted through old Du Pont files for information about leather markets and found reports from 1937 and 1938, when nylon was set to hit the market and Du Pont chemists considered using it for just about everything. Use nylon for leather? Well, you probably could, one report concluded, in upholstery, luggage, handbags, and bookbinding. But not, right away at least, for shoe uppers. True, shoe uppers represented a huge market and good leather got top dollar; so that ought to be the long-term goal. For now, though, the problems were just too great. "Shoe leather is subject to severe and complex stresses, abuse by exposure to temperature extremes and moisture, acids from perspiration, and cleaning methods not under the control of the manufacture." Besides, leather was permeable to water vapor. How were you supposed to mimic that in nylon? All in all, it was "questionable . . . whether a nonfibrous material is suitable in a fundamental way."

But that was 1938. Now, more than a decade later, thanks to Piccard's work, they *had* a fibrous material, or the makings of one, one that might actually breathe like leather. On July 29, 1952, Du Pont lawyers filed a patent application, in the names of John Augustus Piccard and Boynton

Graham, "assignors to E. I. Du Pont de Nemours and Company, Wilmington, Del.," for a "process of forming non-woven porous fibrous synthetic leather sheet." And right there in the patent you could see something of the checkered history of leather substitutes, with all its progress and frustrations:

> Patents relating to methods and techniques of preparing synthetic leather compositions have been granted to inventors as far back as 1850; and even before that time, a great number of attempts have been made to devise a simple and rapid technique of producing synthetic leather compositions. . . . In the early stages of the synthetic leather industry, the main objective was to simulate the general appearance of leather.

Pyroxilin and vinyl-coated fabrics did that, for handbags, briefcases, and the like. But they lacked leather's tear strength, its softness, its ability to transmit water vapor. The new invention, the patent claimed, would achieve all that, its properties making it "equal or superior to the various types of genuine leather."

This patent was the first of many. Six months later Graham and Piccard filed for two more; all three were awarded in late 1955 and early 1956. More than 40 Du Pont patents would be filed by company lawyers before they were through.

There was nothing ambiguous about it: Du Pont was going after leather. Piccard's work had caught the attention of the whole company; everyone wanted in. "Oh, this is obviously a leatherlike film," thought Films, which had just split off from its Rayon department parent. "Oh, thought F&F, this is obviously a new version of Fabrikoid," a Fabrics & Finishes product. That, at any rate, is how Piccard depicted the bandwagon effect years later. For three years, from 1952 to 1955, three Du Pont departments, each with its own history, research personnel, and technical preoccupations, pursued research into the new synthetic leather. Duplication? Waste? By policy and design, the company refrained from curbing what to some might have seemed like it. "Concurrent effort, where two or more departments work toward a common goal, is often found in Du Pont," the company's Corfam chief would explain to a conference on research management in 1964: Competition within the company was as vitalizing as competition from without.

Piccard's work had helped move this chemical colossus toward what would culminate in Corfam. Yet except for the polyester fiber itself, no single ingredient, method, or process from Buffalo in 1949 actually made it into the final product. Piccard had shown that something like it was possible. But what you could do crudely, laboriously, and sloppily in the lab, surely you could find a way to do better, faster, and more reliably. "Various ideas were floated, teams assembled, samples made, patent applications filed and moneys expended," another key Corfam engineer recalled of this period—by one reckoning, 75 man-years' worth of research effort in all.

But in October 1955 the competition was over. The company's guiding arm, its executive committee, awarded the new material to Fabrics & Finishes, maker of Fabrikoid. "They had a naturally greater interest," recalls Joe Rivers. "They knew coated fabrics. They knew shoes." Responsibility for further development would pass to F&F's research group in Newburgh.

You could see why they'd made a fuss over it when it opened. Until then the most prominent feature of this ramshackle industrial site at the edge of Newburgh was the tall brick smokestack that loomed over it. As they had done for four decades, they made Fabrikoid here as well as other coated fabrics, like Fabrilite, a vinyl used for upholstery. But in April 1947, Du Pont announced that on the site's northwest corner, a new research lab would rise. Fifteen months later it opened to much local fanfare—a three-story concrete and brick building, all lab benches, chemical exhaust hoods, tensile test machines, cold rooms, and humidity chambers, to replace the sad trio of corrugated-steel-fronted structures that constituted the lab before. From the third floor you could look out through banks of large windows to the southeast and see the highlands above the Hudson River near the military academy at West Point. The new lab, read the press release when it opened, would help Du Pont "enlarge the scope of its scientific research in a field in which it has been interested for nearly 40 years."

But this was no CalTech or Bell Labs. The Newburgh people, Joe Rivers would say, had all they needed for their own work but lacked much interest in anything else, seeming "happy to work in areas not exactly cutting edge." In 1950, when he and Piccard traveled up here, their porous leatherlike materials had lit no fires. Yet five years later, it was

Newburgh to which Du Pont was funneling new resources, and new people, to carry on the work. Transferred to the new venture were half a dozen people from Du Pont's coated-rubber operation and chemists and engineers who'd worked on other big Du Pont projects, like Teflon, and old line Newburgh hands, like Joseph Lee Hollowell, who would head up the next round of synthetic leather research.

"Bud" Hollowell had started at Du Pont in 1942, working on explosives, had in 1949 come up to Newburgh where, as research supervisor, he'd overseen development of new ways to use PVC. Now, recalls one colleague from those days, John Richards, he was "at the heart of the Corfam thing." He remembers Hollowell as "very bright, conscientious, hard working." And he was not just thinking about leather's breathability.

It may seem that our account thus far has focused on a single key material property—leather's ability to breathe. And certainly this was a quality much sought, and much touted when Corfam was finally introduced. But real materials always embody a trade-off among many qualities. Advertising may ultimately emphasize a car's high fuel mileage, say, or a fabric's washability, but it's never just that one thing you want. All through early development of the material that became Corfam, breathability—its "vapor permeability"—warred with the need for strength, durability, and other technical and aesthetic virtues. One of John Piccard's earliest reports hinted at the problem when he observed that the more porous his experimental materials the weaker; why, the more voids permitting water vapor to escape, it could seem, simplistically, the less material remained to make it strong. Unless you could get a production-line nonwoven material strong enough to serve in a shoe, you might have little to show for your trouble but a fluffy mess of synthetic fiber not much good for anything but stuffing pillows.

In fact the nonwoven fabrics on which John Piccard had made some early progress were in their infancy. They'd begun to show up in the garment industry, in interlinings no one but a seamstress ever saw, and that was only in 1952. Those puffy insulating materials in the walls of your house? They didn't exist. Disposable baby diapers? They wouldn't make real inroads until the early 1960s. At first nonwovens just meant chopping up fibers and finding some way to keep them together. Still ahead was the day in 1955 when Du Pont engineer James Rushton White noticed a veritable cloud of polyethylene fluff piling up outside a pipe in an Experimental Station lab, "like piles of cotton candy," the way Hamilton Fish, who was around at the time, recalls. "Everyone at Du Pont knew it in two days." That was the origin of one approach to nonwovens, flash spinning, which spews out molten plastic, almost explosively, from a high-pressure spigot; it cooled and solidified almost instantly. These days the Association of the Nonwoven Fabrics Industry describes no fewer than half a dozen ways to make nonwovens—wetlaid, drylaid, spunlaid, and so on. But none were part of the industrial vocabulary of 1952 when Bud Hollowell looked to them as the basis for a synthetic leather.

Ordinary felt, which is surprisingly dense and strong, is not always classed as a nonwoven today, but that's about what it is. Early peoples are said to have made felt by washing wool fleece, spreading it out while wet, and beating it, shrinking it, into a mat. It was rediscovered in the 8th century, the story goes, by a Benedictine monk on pilgrimage to Mont St. Michel; the foot-weary fellow is supposed to have stuffed greasy wool into his sandals to ease the pain, at day's end observing that it had become densely matted from the heat, moisture, and pressure of his feet.

But wool fibers have interlocking microscopic scales, like miniature grappling irons, that keep the felt intact. The polyester fibers Hollowell and his group were considering didn't; being smooth, they wouldn't spontaneously form themselves into a mat. Was there, then, some other way to tightly entangle them? For this Hollowell and his group turned to an old, seemingly primitive technology—needlepunching.

Slip a smooth, polished needle—just the needle now, no thread, forget sewing—through a tangle of fibers and it will slide in and out effortlessly, to no effect. But now abrade the needle, or impress it with little barbs or serrations, and it will catch on fibers, pushing them down into the lower reaches of the web, entangling them; do that again and again, with many needles, thousands at a time, and you'd wind up with a three-dimensional jungle of snagged and interlocked fibers; up to a point, the more you needlepunched, the more entangled, and the stronger, stiffer, and denser the final web would be. Needlepunch technology went back to the 1870s, when it was used to make horse blankets out of scrap fabric. The machines were around and Hollowell got his hands on one, with a bank of maybe 5,000 needles, each 3 or 4 inches long, in a 3 × 5 foot array.

Needlepunching, then, got Du Pont engineers part of the way toward the sort of tight dense web, reminiscent of real leather's collagen bundles, they wanted. The other key to density and strength lay in the fiber making up the web in the first place—or rather, what you could do with it. The fiber was just polyester; that had been one spur to John Piccard's work—how to use lots of what Du Pont already made. It wasn't just garden-variety polyester they ended up using, however, but specially drawn polyester. *Drawn?* That is, stretched. Near the end of the fiber-making process it was pulled and stretched, aligning its long molecules into more crystalline form. "Superdrawn," they called it. Once you heated them, however, they shrank. And shrinking them was the whole idea.

Later, in production, they'd start with polyester staple—fibers a couple of inches long, delivered from the plant next door in compressed 500-pound bales, wrapped in cardboard, metal straps keeping them together; in order to retain its shrinkability, it had to be stored in refrigerated storage areas at about 40 degrees Fahrenheit. They'd take the bale apart, load the fiber into a hopper, and feed it into a powerful airstream that deposited a light, loosely gathered layer of material onto a moving belt, a little the way a clothes drier deposits a fine fluff of material onto its lint-catching filter. Then, they'd needlepunch it to make it a little stronger and tighter, and wrap it up onto 10-foot-long cores. At this point the batting was still so light that a three-quarter-inch-thick piece the size of a kitchen chopping block weighed less than half an ounce. Then the web was drawn through three separate baths of warm water, each just the slightest bit warmer, each controlled to within a tenth of a degree. The fibers shrank and with them the web as a whole. Its width was reduced from 10 feet to 6. Its density was now 10 times what it had been before, the entangled polyester batting now something closer to the fleshy underside of leather.

Needlepunching. Web shrinkage. The third key technological innovation developed in the 10 years after Piccard's initial work came in 1956. Hollowell's group had learned of the textile fibers department's work with polyurethanes, a class of rubberlike polymers, which would go into the stretchy fiber that became Lycra; Du Pont had been working on them on and off since the 1930s. Hollowell ordered some to study.

A sample arrived, dissolved in dimethylformamide, or DMF, an organic solvent introduced commercially a few years before. A member of his team, Robert Johnston, prepared a film to dry out over night that they could work with the following morning. Newburgh ain't the Deep South; it's more wintry snow drifts than New Orleans swelter. Still, it sometimes endures summer hot spells where the air hangs heavy with humidity. It was on just such a day, probably in late summer 1956, that Johnston laid out his film on a glass plate.

The next morning the polyurethane should have been transparent but wasn't; it was smooth, elastic, and white, its milky whiteness due to an irregular network of tiny holes, visible under the microscope. The solvent, DMF, was hygroscopic; it had an affinity for water, in this case the night's moist air. And when it drew in water vapor, the polyurethane coagulated—coagulated, like curdled milk or clotted blood—leaving soft porous masses, tiny voids running through the material, just the sort of permeable, microporous structure Piccard and everyone else had wanted all along.

Piccard had gotten porosity by dissolving out impregnated fibers with solvent. The patent literature over the coming years would describe many other ways to do it. You could imbue a material with salt, wash it out, dissolving the salt, leaving pores behind. You could use blowing agents that liberated gas when heated. You could swell the structural fibers in a material by steaming it, then dry it out, leaving annular spaces around the fibers. The way that led to Corfam, however, lay in the patent filed early the following year by Johnston, Ron Moltenbrey, and E. K. Holden, a Yale-trained chemical engineer who'd joined Du Pont in 1948: After the polyurethane polymer "has been applied to a substrate, the coated substrate is exposed to an atmosphere containing water vapor. The hygroscopic solvent in the coating absorbs the water vapor thereby coagulating the polymer and depositing a microporous coating of the polymer onto the substrate." It was much trickier than the patent made out, of course, and controlling the process proved difficult; one review of it, based on two decades' experience, lists no fewer than 16 factors—viscosity of the solution, control of trapped air, DMF concentration, temperature of coagulation bath, and so on—that could influence quality. But the essential step was there.

By 1956, Du Pont engineers and chemists had prepared a single sheet of material, about a foot and a half square, made by hand over a period of weeks, that was a crude precursor of what would become Corfam.

"The cost of that piece was so astronomical nobody had the courage to calculate it."

ii. Nylon, Again

"Occasional samples with pliability, strength, porosity, and leatherlike fold had been achieved." That's how John Richards, an organic chemist out of Princeton who had become assistant director of the Fabrics & Finishes research division, later described the status of the Corfam substrate, its underlying fibrous material, in late 1957. At that time the decision was made to install at Newburgh a semiworks, a miniature experimental factory, using mostly used machinery, for the manufacture, in 25-yard lengths, of the new material. By late 1958, Du Pont was making narrow rolls of it and fashioning it into test shoes.

In 1959, Du Pont decided to invest in a pilot plant with about 10 times the capacity of the semiworks. This took about a year to get up to speed and, as Richards remembered it, meant throwing together the machinery for one key process with "plywood, baling wire, and surplus rolls and drives." Ultimately it would reach a capacity of about a quarter million square feet per month, make Corfam for thousands of pairs of shoes for consumer testing and supply shoe manufacturers with material for their first production runs.

In October 1962 the announcement was made that a full-scale plant would go up near the site of three other Du Pont plants in Old Hickory, Tennessee, not far from Nashville. It would have a capacity of about 30 million square feet per year.

The following month a poromeric products division was created within Du Pont's Fabrics & Finishes department. Bill Lawson was named head.

In 1963, Corfam was introduced to the world.

During the whole period of Corfam's ascendancy—from early whispers about it in the late 1950s, to its introduction, and for several years after that—only one note sounded by Du Pont could be heard: Corfam was an implacable force, its success inevitable. Of course, you might expect a new product to enjoy a flurry of corporate proud papa huzzahs. Corfam, though, was extraordinary for the incessant drumbeat of publicity the company generated for it. "Du Pont's lavish introduction of Corfam was like a society debut." This was *Chemical Week*'s take, looking back from 1967, and they had it about right. Indeed, it was actually staged over seven months, to three distinct audiences—shoe manufacturers, retailers, and the press.

At the first, in the Baroque Room of the Plaza Hotel in New York, Du Pont introduced to one another Corfam's "charter" shoe manufacturers: Nunn-Bush, Stetson, Palizzio, Mademoiselle, and close to two dozen others. A two-and-a-half-hour lunch was held for the women's shoe manufacturers, a dinner meeting for the men's. Amid displays showing the historical development of the shoe industry, they heard about the hundreds of newspaper ads and radio commercials that would tell America about Corfam. They learned of television commercials set for *Du Pont Show of the Week* and got swatch books of Corfam colors and textures. The way marketing chief Charlie Lynch made it sound later, it was a kind of bonding. "We wanted to give them a chance to look at our support program and especially at each other—to know they were in good company." They and their fellow innovators, he assured them, stood to make "the greatest contribution a small group has ever made to the overall good of the shoe industry in our time." The day was June 23, 1963, three days before President Kennedy's *Ich bin ein Berliner* speech at the Berlin Wall.

Four months later came the National Shoe Fair in Chicago.

WORLD PREMIERE
DU PONT PRESENTS
"THE STEP INTO TOMORROW"
INTRODUCING
CORFAM
THE BREATHABLE, MAN-MADE SHOE UPPER MATERIAL

So read the company's invitation, in an assortment of flowing typefaces, in a four-page trade industry ad. Six theatrically lit lobbies in the Normandie Lounge of the Conrad Hilton hotel, 3,500 square feet in all, were given over to showing off the handiwork of the top charter manufacturers, to taking Corfam through a succession of shoemaking operations, to introducing Du Pont's vision of the modern shoe store.

SEE THE PERFORMANCE OF CORFAM
IN DRAMATIC DEMONSTRATIONS THAT SHOW YOU THE FACTS
BEHIND THE EXCITING, SPELLBINDING
FLAIR OF CORFAM.

Then, on the eve of the advertising campaign set to begin on January 26—it was two months after the Kennedy assassination and Lyndon Johnson was president—came the press introduction at the New York Hilton. There reporters watched a film showing a Corfam-shod model exiting a taxi in front of a hotel, being whisked through its lobby, onto an escalator. And then, right on cue—now live of course—she stepped before them into the hall.

The advertising was memorable even by today's hype-drenched standards. Du Pont advertised, and even advertised its future advertising.

IF YOU WERE INTRODUCING DU PONT'S NEW CORFAM
WHICH ADVERTISING CAMPAIGN WOULD YOU CHOOSE?

Color pages in *Vogue*? Full-page newspaper ads in 20 key markets? FM? TV, on *Du Pont Show of the Week*? Eight full-color pages in *Harper's Bazaar*? Color spreads in the *New Yorker*? "Which campaign," Du Pont asked readers of *Boot and Shoe Recorder*, would they choose? An arrow pointed them to the next page, where the answer awaited them in a 3-inch font.

ALL OF THEM.

"Forget everything you ever knew about shoes," said a flowery four-page color spread in *Vogue* showing off a black pump by I. Miller, a navy sling pump by De Liso, and eight other name-brand models arrayed appealingly on a red mat of Corfam. "Step into tomorrow with a remarkable new material from Du Pont."

Around this time Du Pont ran a series of training events for retail store salespeople at department stores across the country. In Brooklyn it was in the fourth-floor restaurant of Abraham & Straus's downtown store, warmed by coffee and danish. In New Orleans it was at D. H. Holmes on Canal Street. Typically, the events ran half an hour and featured a short color film, *Step Into Tomorrow*, extolling Corfam's fashion and technical pluses, especially its breathability. Fact cards and booklets were distributed, questions addressed, free pen-and-pencil sets dispensed. The big draw, though, was usually the mud box, thick with gloppy mud brought from Wilmington that Dupont's trainers would spread over a Corfam shoe and then serenely wipe off. "They couldn't believe it till they tried it for themselves," recorded the New Orleans rep.

These presentations, which reached more than 3,000 salespeople, went over big; we know this because Du Pont surveyed them and gathered the numbers: 66 percent were "very enthusiastic," 26 percent modestly so, while 1 percent indicated "no enthusiasm." As Corfam went before the world in early 1964, there was no room in the heavens for anything but blue skies and boundless expectations. "We don't have the limitations of preconceived ideas," Charlie Lynch told a reporter for a trade magazine before the Chicago opening, referring to design possibilities for Corfam. "We will seek to move in fresh directions, initiate new ideas. We won't accept the concept that it can't be done. We're ceaselessly confronting ourselves with the question: Why can't it be done?"

A Du Pont presentation purporting to offer "A Research to Reality Case History" of Corfam, probably in early 1966, lamented "the inadequacies of leather" and reminded listeners of childhood admonitions to keep one's leather shoes out of the rain lest they be ruined. "The greatest compliment that can be paid to leather is that it has been able to stand so long without serious challenges." Not for long, though. Corfam, Bill Lawson said at its introduction, would find use in upholstery for car and home. Likewise "draperies and wall coverings that run the gamut in pattern design from aardvark to zebra." Bejeweled evening slippers. Corfam sport jackets "so like suede" you could tell they weren't only because you'd be able to run them through the wash.

By the time Corfam burst on America, Du Pont was thoroughly pleased with itself, delighted with its new product, and eager to tell anyone who'd listen all about it. About the 20,000 pairs of test shoes it had made with it. About its breathability. About its tenth-of-a-degree temperature control. About how it wasn't a coated fabric, wasn't plastic, wasn't leather, but would "take its place, on its merits, shoulder to shoulder with the natural material."

And how could the leaders of the charge not feel it themselves? For young executives like Charlie Lynch, Corfam was their professional trial by fire. Du Pont people weren't shoe people, Lynch tells the story, and had much to learn about the industry. Early on, a shoe manufacturer asked him of the new material, "How does it last?" Oh, he replied, it lasted a long time, wore very well, thank you. Of course, that's not what the shoe manufacturer wanted to know, but rather how Du Pont's material endured the lasting process. "We didn't have a clue," Lynch says. But after that embarrassment, "We came away determined to get experience within the industry." They went into factories, saw leather being cut and shoes being made, visited shoe stores, he and his colleagues putting in turns as salespeople on the floor. "It was a helluva experience." By the time Corfam was out in the market, Lynch could brag about Du Pont's close links to the shoe industry. "Most technical service groups, even at Du Pont, are made up of chemists," he said. "Ours consists of shoemakers. Not one has a college degree, but each is an expert." They were proud, proud of Corfam, perhaps scarcely able to imagine for it anything but bright prospects and bountiful success.

Seven months after its introduction, in August, a Du Pont ad styled as a "status report" told retailers unable to get their hands on Corfam shoes that its extraordinary popularity was to blame. "Even in their most optimistic dreams," it asserted, Du Pont forecasters could not "foresee that Corfam would be snapped up as quickly as it was."

> Corfam actually *leaped* from introduction to full consumer acceptance.
> No gradual acceptance.
> No "wait-and-see" attitudes.
> No struggle to overcome old habits.

An "instant success." Hence the wait.

The Old Hickory plant was nearing completion, but Du Pont would not rush it. So please, "We ask your patience," because the "highly secret" process for making Corfam required "a manufacturing plant that's completely different from every other factory in the world." And here Du Pont's copywriter appended one final theme, one that had sounded through the Corfam project like a mantra from the start. The factory was different, yes, "just as the first nylon plant was unique, and took extra care to build."

"It is obvious that the new material is a very special kind of wonder," wrote Lawrence Lessing of Corfam in a rhapsodic early *Fortune* article. "In scale and significance it may well be another nylon." Corfam was the next nylon. Everyone said so. The Corfam factory had problems, but so had the nylon factory. Corfam was a stunning innovation, like nylon. Corfam might mean "the end of Tony the bootblack," make shoe polishing obsolete—"just as nylon," an early news report quoted a Du Pont spokesman as saying, had "obsolesced the darning of socks." "Remember what nylon did for undies?" a Baltimore shoe store asked in a local ad. Well, Corfam was "the newest 'how-did-I-ever-live-without-it' material for shoes." All through the introductory period and beyond, the invocations of nylon droned on. Into a corner of its "status report" ad, the company actually tucked a facsimile of an ad from 25 years before, in which "Du Pont Announces for the World of Tomorrow . . . a new word and a new material: Nylon." Nylon was Du Pont, Du Pont was nylon, and now Corfam would carry its legacy unto the next generation.

What Wallace Carothers's man had cooked up in the lab at the Experimental Station had sparked an industry. Enshrined in the public mind by postwar images of women queued up for blocks to buy stockings—more than 60 million pairs the first year—nylon became indispensable to modern life: toothbrushes, tennis racquet strings, surgical sutures, parachutes, fishing line, machine bearings, tire cord. Along the way it made Du Pont a fortune and linked its name forever with benign visions of industrial research. From its labs sprang a succession of new synthetic materials: "Better Things for Better Living Through Chemistry." Better, stronger, lighter. Research and development successes not necessarily the product of solitary genius but of a whole corporate culture. The symbol of that culture was nylon.

Now, a quarter century later, Corfam was being pitched as the next nylon—first by Du Pont, then by everyone else who'd imbibed its message, which was pretty much everybody. "Corfam was not just another product," recalls Hamilton Fish. "Corfam was going to be the next nylon. A lot depended on it." Parallels between the two materials were made to seem downright uncanny. Nylon and Corfam were both synthetic replacements for natural materials, silk and leather, respectively. In early tests of nylon, Du Pont security was so tight that afterward every scrap was gathered and weighed to ensure that no fiber was missing; identical security precautions, down to rounding up the scraps, were taken with Corfam. Stockings and lingerie were distributed to the wives of men working on nylon; shoes of Corfam went to the families of those working on it. Nylon, the name, was the product of a Du Pont committee's protracted search; so was Corfam, though a computer was enlisted to generate possibilities. Nylon was introduced at the 1939 New York World's Fair with the slogan, "We Enter the World of Tomorrow." Corfam was introduced, 25 years later almost to the day, with the slogan, "Step Into Tomorrow"; a few months later it was featured at the 1964 New York World's Fair, where at the Du Pont show, "The Wonderful World of Chemistry," a cartoon told "the animated story of chemistry's evolution from the caveman to Corfam poromeric material."

One other curious parallel emerged during this period. In the March 1964 issue of *Reader's Digest* appeared a "Report to Consumers" about Corfam that verged on outright endorsement: "It looks like leather, feels like leather, wears like leather, and *breathes* like leather" and "seems designed to walk right down Main Street."

The tanners were furious. The director of Leather Industries of America, Mel Salzman, pulled its ads from the *Digest* and fired off an angry letter to its publisher. He'd never, he said, seen such "blatant editorial endorsement of a commercial product," one that took "untested and unproven claims" and presented them as fact. Its author asserted that Corfam's structure was "basically the structure of leather." No, wrote Salzman, Corfam "does not begin to approximate the natural fibre structure of leather." The author claimed it wasn't a plastic, when most people in and out of the shoe business regarded it as precisely that. He accepted without question Du Pont's promotional arguments, thus serving as company mouthpiece. His information could have come from "only one source, the manufacturer of Corfam: Du Pont."

Indeed, the author of the *Digest* article, which came out just as Corfam was being introduced and ads were appearing in magazines around the country, wrote that he'd been the only journalist allowed inside the pilot plant. He'd gotten the chance to see "Delaware River mud" rubbed into a Corfam shoe that, cleaned with soap and water, "emerged spotless." He'd recorded, without demur, Du Pont's claim that the Corfam-making process was "as sophisticated as the extraction of fissionable material" for the atomic bomb. His name was Lacy Donnell Wharton, Jr., but he went under the byline Don Wharton. And two decades earlier he'd written a similarly glowing piece about nylon. Appearing in *Textile World*, January 1940, it was entitled "Nylon—A Triumph of Research."

Within months of its introduction, with Corfam's success seemingly assured, three Du Pont executives—Bill Lawson, Charles Lynch, and Jack

Richards—went down to White Sulphur Springs, West Virginia, for the spring meeting of the Industrial Research Institute. And already the "case study" of Corfam's development they presented there had something of the triumphant air that would mark Corfam's commercial life. When it appeared the following year in *Research Management*, the editor, introducing the issue, wrote that Corfam had

> surmounted the formidable barriers that always confront efforts to establish a completely new product in an entrenched market. This paper reveals what can be accomplished when superb resources are committed toward definite objectives with masterly skill.

Soon after its introduction, *New York Times* reporters bought Corfam men's and women's shoes and some months later reported on their experience with them. The male reporter pronounced his shoes "rugged and durable," with fit, shape retention, and breathability all satisfactory. In the same article one of the chief proponents of Corfam among the shoe manufacturers, Maxey Jarman of Genesco, was quoted as saying that "eventually" Corfam would hold a sizable proportion of the shoe market, business worth $300 million a year or more. All was proceeding apace.

At the time of Corfam's introduction in January 1964, Du Pont issued what would today be called an FAQ, a seven-page list of frequently asked questions. One of them read:

> Q. What disadvantages are there in the material?
> A. All shortcomings encountered in the development of Corfam have been eliminated.

iii. Hot Feet

But now it was 1969, and there seemed to be some sort of problem.

Du Pont had commissioned a national survey of about 1,000 consumers. About 18 percent had bought a pair of Corfam shoes, three-quarters of them saying they might buy another. Of the much larger number who had never bought Corfam, about 1 in 5 said they simply preferred leather or didn't like synthetics. All told, this wasn't too bad.

What was bad, very bad, were the results of another survey, this one of retail shoe salespeople, 460 of them. Only 1 in 10 said Corfam was easier to sell than leather, 4 in 10 that it was harder. Corfam shoes, they said, were hot. They burned your feet. They didn't breathe. They didn't stretch.

They were stiff and hurt your feet. They were hard to fit in the first place. These were substantive, damning complaints. And they were made by people who knew shoes, knew feet, knew fit, knew their customers. In a report citing this study submitted to Du Pont's executive committee in November 1969, entitled "Whither Corfam?," Fabrics & Finishes general manager Richard Heckert declared: "The foremost product problem is comfort."

One day some years before, Bob Wilson's dad handed him a pair of new shoes, saying, "Hey, they're giving these out to test. See how you like them." They were cordovan-toned wingtips, made of Corfam, and his dad was high on them. "They'll be the next thing to replacing leather in shoes," he recalls his father saying. Bob Wilson had grown up in Newburgh, New York. His father worked for Fabrikoid and was now the division's export sales manager. Young Wilson, then in his early 20s, dutifully wore the new shoes—and suffered for it. "I was willing to give Dad the benefit of the doubt," he says, "but I'll tell you they were the most god-awful uncomfortable shoes I'd ever worn in my life."

How could this be? How could it be when foot comfort depended on breathability, around which Du Pont had designed Corfam and which had been bruited about from the beginning as its cardinal virtue? Did Corfam not breathe?

In 1967 two researchers reported to the British Leather Manufacturers Research Association that Corfam's water vapor permeability was about one-third that of leather—but vastly better than that of vinyl, which was entirely impermeable. Three years later the big British testing organization, the Shoe and Allied Trades Research Association, tested Corfam against leather and other poromerics and found that, whereas a square centimeter of ordinary side leather transmitted 3.0 milligrams of water vapor per hour, and calfskin 6.9, Corfam came in at 2.5. Other results, by American testing labs, were similar.

So it *did* breathe. Not as well as leather, maybe, but well enough. Du Pont might be taking a little poetic license with the facts, but it wasn't lying.

Well, then, how could these troubling reports of "hot feet" be true? They had tested Corfam, hadn't they, on real people? For strength, durability, wear, and comfort? It was just such field testing that Du Pont cited in every press release and every ad. The company loved talking about it.

When they asserted that Corfam had no shortcomings, it was because, they said, any early problems had been picked up through shoes wear-tested "with the express purpose and object of discovering and overcoming any weaknesses in the material."

Du Pont's consumer testing wasn't unique to Corfam. At one point in 1961, *Du Pont Magazine* reported that some 3,000 volunteers "from Maine to Miami" were wear-testing clothes made with its synthetic fibers. They included "pretty girls" in Lycra swimsuits at Barry College in Florida and brawny butchers wearing insulated underwear in Philadelphia. But Corfam testing was on another scale entirely. The number of pairs of Corfam shoes Du Pont said it tested depended a little on when you heard it, but the figure settled on in the end was a nice round 20,000. Nurses in hospitals. Children in orphanages. Cops on the streets of Newburgh. One canned publicity photo showed a uniformed Newburgh policeman, his Corfam-shod foot resting on a fire hydrant, as a Du Pont man, clipboard in hand, listens to him earnestly. Another shot pictured a white-coated technician—actually, he's the same model used for the Newburgh photo—squatting over what looks like an airplane hangar-scaled sea of shoes stretching to the horizon. "Massive line-up of worn shoes," says the caption, "documents extensive field testing of 'Corfam' poromeric material before it was introduced to the public."

In January 1964, a 21-year-old U.S. Marine Corps corporal, Max W. Carlson, saw an ad for Corfam shoes and wrote Du Pont suggesting that he walk cross-country in them, at Du Pont's expense. Sure, Du Pont replied, with what must have been a big corporate grin. On Carlson's 2,000-mile hike, from the Marine air station in El Toro, California, to back home in Minnesota, he went through half a dozen pairs of leather soles and 10 pairs of heels. But the Corfam uppers did fine. "Only two blocks to go. One block," *Du Pont Magazine* breathlessly reported in late 1965, beside pictures of the young Marine, standing tall, traipsing across the infinite American West. "There it was. . . . The Minneapolis city line. Max (and 'Corfam') had made it."

Virtually everything Du Pont issued during these years was by way of proof that Corfam was equal to, or better than, or fit as well as, or lasted longer than, or made as good an impression as leather. "In pursuing our marketing effort," Charles Lynch said at the White Sulphur Springs conference, "we have taken a unique new material into an arena steeped with tradition and emotion." And while the drumbeat of good news was, of course, mostly for the sake of Du Pont customers, it no doubt promoted morale in-house as well, discouraging any arrant thought that anything might not be just fine.

But there was plenty that was not fine. And hints had been there all along. Consider, for example, the St. Luke's study.

Du Pont supplied the material. Du Pont hired the fitter from a local department store. Du Pont supported the researchers, Edward C. Meldman and Arthur E. Helfand. The two podiatrists went into St. Luke's and Children's Medical Center in Philadelphia, recruited nurses, and told them they were being issued shoes—nurse's classic white oxfords—"to test a new way to keep white shoes clean without work." This was untrue. The test was really to see if Corfam shoes were comfortable and good for your feet. The results? No pathologies. No allergies. Everything OK. Meldman and Helfand reported their results in an article in the *Journal of the American Podiatry Association.* Corfam, they declared, "is comfortable and compatible with good foot health, when used as a footwear upper material under conditions otherwise also compatible with efficient foot function"; this reads much like the early, much-quoted endorsement of fluoride toothpastes by the American Dental Association. Soon Du Pont was running ads like this:

NOW YOU CAN TELL YOUR CUSTOMERS:
CORFAM HAS BEEN
TESTED AND APPROVED
BY THE AMERICAN
PODIATRY ASSOCIATION

There was only one problem. Over the course of the St. Luke's study, the nurses would check in every two months and each time the number participating in the study dropped. From the originally selected 106 to 98 after two months, to 89 after four, and 82 after six, at which point the study was over. It wasn't because the researchers had trouble tracking down their subjects. The nurses worked at the hospital; they were right there. So that wasn't it. No, they had simply been eliminated from the study because of what the researchers termed "fit failure."

Each nurse had been asked for her shoe size, then supplied with shoes. But two months later, eight of them said that, well, their shoes didn't fit after all. At subsequent check-ins, more of them said the same thing. The project staff had anticipated some fit problems, since the nurses hadn't actually been fitted, merely supplied shoes in their accustomed sizes. But why should the number of dropouts steadily increase? Meldman and Helfand asserted that the problem was "not related to the upper material"; I report to you just what they said. What, then, *was* it related to? It "appears to reflect adversely on the reliability of the subject's own shoe size preference as to 'proper fit.'"

The subjects, they as much as said, were to blame. *Why, they didn't know their own shoe sizes!*

In retrospect the explanation seems plain enough. The subjects knew their shoe sizes all right. They also "knew" that their shoes would adjust to their feet. Leave the store with shoes a bit tight and, they'd learned over the years, their shoes would give a little, sop up moisture, yield as living skin does, ultimately come to fit like "an old pair of shoes."

But their Corfam shoes never did. Rather, they stretched to fit feet around which they might be uncomfortably tight, then shrank back to their original size once removed; wearing Corfam, many would later say, was like breaking in a new pair of shoes every day. Some years ago an American company started making shoes from elk and bison leather, boasting of "Old Shoe Comfort . . . Right From the Start." Well, for some wearers of Corfam shoes, it was more like "New Shoe Discomfort . . . From the Start and Always."

Corfam's resistance to stretch had originally been touted as a selling point. At his introductory press conference in 1963, Bill Lawson said Corfam shoes would "gently resist stretching out of shape and becoming distorted or sloppy on the foot." No break-in required. "Put on your shoes in the morning, and they will have the same shape as when they came out of the shoe box, perhaps months ago." They wouldn't sag, gap, bulge, or stretch. Of course, Lawson added, proper fit was essential. "You cannot depend upon torturing them into shape later."

But the dark side of Corfam's shape-holding propensity surfaced early on, in late 1964, when Consumers Union, publisher of *Consumer Reports*, weighed in on the new material:

> Perhaps the most important bit of information to come out of CU's experience with the Corfam shoes concerns fit. Although Corfam is pliable,

> it does *not* stretch or eventually mold itself to the foot as leather does. If a Corfam shoe pinches at first, it will continue pinching. So, proper fit at the time of purchase is even more important than it is with a pair of leather shoes.

Consumer Bulletin, too, noticed Corfam's shape retention and also pointed out "that the synthetic material will not develop the gradual local changes of shape that permit it to conform to the shape of the individual foot, as leather does."

Sure enough, foot comfort complaints began coming in from retail accounts. Hot feet. Sweating. Pain. Discomfort. Not from all wearers of Corfam shoes. But many. Too many.

Du Pont had at least an inkling of the problem from the beginning. At its sales training meetings in 1964, the second most frequently asked question was, "Is there a fitting problem with Corfam shoes?" A point driven home at these sessions, according to an article in *Chemical Week* in 1967, was: "Corfam shoes must fit perfectly because they do not 'give.'" But the message never quite got across, in part because, as one Du Pont competitor told an interviewer that year, retail store fitters were used to selling "the size they happen to have in stock or the size the customer requests"; after all, with leather shoes that was usually good enough.

"With impeccable logic," the way someone put it later, "Du Pont officials pointed out that [fit] would be no problem if customers moved up to a slightly larger size. But this rational observation bumped into the irrational psychological fact that not many men, and fewer women, like to acknowledge that they have big feet." An old Corfam hand reports that one manufacturer took to selling size 10 shoes as $9^1/_2$. Meanwhile, in advertising and promotion, Du Pont reemphasized the importance of correct fit. And a new, presumably more conformable Corfam was supposed to be in the works, initially aimed at nurses and other wearers of white "duty shoes." But perceptions were taking hold: "As comfortable as an old pair of shoes" would never be said of Corfam.

A draft talk on Corfam and foot health prepared for a German medical association meeting in 1967 said Du Pont's goal was to ensure "that Corfam, the material of the future, would offer the wearer the same foot hygiene characteristics as had the traditional material, leather, for so many centuries." But if foot hygiene meant comfortably adapting to the projections and protuberances, the sweating and swelling, of defiantly individual human feet, it failed.

iv. Hard Days at Old Hickory

Meanwhile, Du Pont was having trouble making the stuff.

In his slide show at the Experimental Station, Piccard's colleague, J. E. Evans, had used simple schematic drawings to sketch the manufacturing of leatherlike nonwovens, little boxes representing calendering rolls, baths, dryers. But "to avoid any misunderstanding," he added, "this process has not as yet been tried out in this integrated form. This is the target toward which we are shooting. . . ." Evans gave the talk in September 1950. Twenty years later Du Pont was still shooting, and not always hitting.

Like Evans's slide show, the basic recipe for Corfam could seem straightforward enough: From polyester staple (fiber cut to length), form a light, fluffy web, needlepunch it, shrink it, impregnate it with polyurethane binder, coagulate it; that got you the "substrate," the sturdy substance of the synthetic leather. Now, coat the substrate with polyurethane, coagulate it to make it porous, emboss with a leather grain. For some uses a thin woven cloth interlayer fit between substrate and the polyurethane coating. That was about it, reduced to a few words. But getting it to work, even as well as Du Pont ultimately did, scaled up to a big plant, was plagued with problems; making Corfam was *never* easy and *never* economical. "Du Pont is selling all it makes; it is straining to meet demand," a trade journal reported early in 1965. True enough, but that assessment reflected as much the difficulties of making Corfam as the demand for it.

The difficulties started in Newburgh, with the semiworks, the miniature plant intended to supply material enough to make shoes for testing and to perfect manufacturing operations. As Jack Richards recalled a little later, "The equipment as installed was more helpful in defining problems than in solving them. It turned out that we had underestimated the level of engineering sophistication needed to develop a really workable process."

The problems were still there in the pilot plant, at 10 times the production volume. One Du Pont veteran remembers visiting Newburgh and seeing "junk piled up in the corner. They made a lot of stuff that wasn't saleable. They were trying to figure out how to make it good" and not yet succeeding. They were making footwide lengths—this was probably around 1962—and "things would happen at the edges or the middle that would make it totally worthless." Soon after Corfam was introduced, a reporter said of the Newburgh plant that it "hummed" with chemists and engineers, scores of them, "split into specialized teams . . . working to refine the qualities of Corfam poromeric material." That may have been a discreet way of alluding, as part of *Boot and Shoe Recorder*'s almost wholly laudatory account, to Corfam's production problems.

Take web shrinkage, a key Corfam-making step. At the beginning, especially in the semiworks and the pilot plant, Richards admitted, they'd had "no concept" about how to do it. The principle, recall, was simplicity itself: Heat the polyester web and watch it shrink. But the heating had to be precisely controlled. "If we put it in hot water to shrink," recalls Richards, "it puckered all to hell." They tried using a succession of water baths. The contraption sat in the middle of the plant like a big boat or barge, maybe 60 feet long, its three compartments filled with water of successively hotter temperatures. And there he was, John C. Richards, director of the Newburgh Research Laboratory of the Fabrics & Finishes department of E. I. du Pont de Nemours at one end, and another high-priced colleague, Bob Fry, at the other end, handcranking the fragile web through, the stuff fairly "creeping along," a few painful inches a second.

Of course, that was Newburgh, the pilot plant; that's where you were *supposed* to work out snags like that. But to listen to some of the hundreds of Du Ponters who worked at Old Hickory, including engineers and chemists transferred en masse from Newburgh, things were never that much better down there. Even after more than five years, says Ron Moltenbrey, who'd joined Du Pont after graduating from Worcester Poly with a chemical engineering degree and worked on Corfam from 1955 into the 1970s, "We were running the world's largest pilot plant."

The wife of a Corfam executive remembers being surprised on her one visit to the plant—"appalled" is actually the word she used—that making Corfam wasn't some glorious, robotically efficient operation, that men in aprons had to, well, *move things around.* Awesomely automatic industrial processes, one stage leading seamlessly into the next, untouched by human hands? With Corfam, it never happened. At Newburgh they'd been making Corfam on a small scale, a quarter million square feet per month, and at Old Hickory were supposed to make 10 or 20 times that much. But robotic it wasn't. Richards, who became assistant director of the F&F research division, remembers 15 distinct production steps at Newburgh, each requiring an unwinding and winding of material, or "doff." The word was imported from the textile business and implied a punctuation to each step; at Newburgh, Corfam required 15 doffs. "It's a nice concept to have all those 15 steps running together," says Richards, but hard to achieve since, for example, some processes are fast, others slow. "All those wind-ups and unwinds—you can't just couple them together in one long string."

For Bud Hollowell, who like so many of the others, had come down to Old Hickory, that unwieldy stringing together fatally undermined Corfam's chances. "The process was extremely complicated with many steps, such that even with high yield for each step per se the overall yield was low." What he meant can be likened to one of those simple probability problems where to figure the likelihood of a particular outcome, you multiply the likelihood of each step in the train of steps upon which it depends. The less likely each one is, and the more of them there are, the less chance it will ever materialize. So it was, roughly speaking, with Corfam: Any yard of it actually getting out the door depended on numerous processes, each difficult and uncertain, each working just right.

One problem they encountered—not more troubling than any other, just one of many—was that the needles used in needlepunching broke. They installed metal detectors—electromagnetic coils which, detecting a broken needle, would issue an alarm, prompting a machine operator to fish it out by hand from the web; miss a needle and it could wind up in the final product. In one series of almost 200 lengths of web, the making of two in three was slowed by at least one broken needle.

Then there were the chunks of polymer in the dip bath. Once needlepunched and shrunk, the web was impregnated with polyurethane. The web entered a sealed structure, recalls John Noble, then a young field engineer, where it floated in heated polymer in a long pan, maybe 6 inches deep. During its trip through—you could look in through windows along the side—the web wicked up the polymer, any excess material being stripped off at the far end. From here the impregnated web dropped down into the coagulation bath. This was the process, its origins in that humid summer night of 1956, that gave Du Pont its claim of a million pores per square inch. But controlling it had required much refinement, including several patented innovations; it was never less than complicated. At one point, the way Noble describes it today, the web was wrapped up onto idler rolls, then dipped into the fresh water bath. But the web at this point was not always as nice and evenly impregnated as it was supposed to be and often had floating chunks of plastic hanging off the edge. This would not do; an operator was assigned to remove them by hand.

"Red ink flowed out of the Old Hickory plant about as fast as Corfam," an article in *Fortune* reported later. Broken needles and arrant globs of plastic were part of the reason. Even after almost 20 years of trying to make it—in the lab, in the semiworks, in the pilot plant, and finally at Old Hickory—they couldn't make it right and couldn't make it cheap, at least not at the same time.

"We had problems producing it as efficiently as we'd thought," says Charles Lynch, mildly.

"You couldn't make it nonshiny," laments a former visitor to the Newburgh works, referring to fashion changes to which Corfam couldn't adapt.

"We can't make what they want to sell," Ruth Holden remembers her husband, Old Hickory plant manager Ellsworth K. Holden, laden with scraps of defective Corfam, grumbling to her at the end of a grueling day.

Ironically, what made Corfam's burden heavier was Du Pont's own claims of superiority to "conventional materials," meaning leather. Uniform rolls of uniform material: that's what Corfam promised shoe manufacturers. "Cutting this material is like cutting cheese. It doesn't have any defects, so you can cut many layers at one time," an early champion of Corfam among the shoe manufacturers was quoted as saying. But that was 1962; Corfam wasn't even in full production then. And when it was, at Old Hickory, it didn't always come out so uniform. Splotches. Misbegotten holes. Material 25 inches wide one day, 32 the next. "They could not make the stuff the same way yesterday, today, and tomorrow," says Thomas J. Leonard, then a Du Pont regional sales manager. He feels that the shoe manufacturers, used to working around the maddening vagaries of natural leather, could have managed. "But that was not acceptable to manufacturing, who were hearing from us all the time this doctrine of uniformity."

The problems were there at Newburgh at the beginning and at Old Hickory at the end. Libby Fay's husband, Robert Fay, came to Newburgh around 1952, soon got transferred to the Corfam project as director of the

Newburgh research laboratory, and ultimately went down to Old Hickory. He and his colleagues were excited by the work, she recalls, "trying to see if they could make a decent material out of it." They could, sometimes. But then "they'd get the runs set up, all the kinks worked out, they'd start the rollers, and then something would go wrong," with bubbles and streaks and who knows what else. And all the while they grappled with bubbles and streaks, they were trying to respond to the shifting winds of fashion for new colors, textures, and styles—winds that their natural competitor, leather, seemed to handle with aplomb.

v. Dirge

Meanwhile, reports came in of mysterious flex failures, surface tears from constant flexing of the vamp, especially in men's shoes, that no one seemed able to explain. And while Corfam was scuff resistant, when it did scuff it did so irredeemably; the shoe couldn't be saved by polish and elbow grease, as leather shoes could, and often had to be discarded.

Then there was the matter of softness. It was the late 1960s now, hippies were everywhere, hair was long, flowing Indian cotton bedspreads reached across the nation. Soft was in. Materials didn't gleam; they softly glowed. Leather was soft or could be; leather with a hard sheen was out, too, while suede and warm waxy leather were in. The tanners, energized by the Corfam threat, seemed able to supply whatever the market wanted. "The leather industry," lamented Bill Lawson in 1972, "responded to the poromeric threat by producing and promoting very soft, glovelike leather" tuned to the times. Corfam, on the other hand, was always pretty much the same—smooth, firm, and unyielding.

The year 1969 was a bad one for Poromeric Products' western region, according to a brutally frank marketing report issued early the following year. Shoes in general were weak; traditional men's shoes like brogues, which used typical Corfam finishes, were especially so. Corfam sales were down 44 percent from the previous year. "The year 1969 saw the beginning of the decade of softness," its author observed. "Everything had to be soft, comfortable and conformable. Unfortunately, Corfam was none of these. . . . There developed a stigma about Corfam because it wasn't soft and wasn't 'fashion-right' and therefore it wasn't considered really appropriate for anything." *Ooh,* that hurt. Shoes of Corfam were like those of five years before; they had the hard sheen of a soldier's spit-shined dress shoes, not the warm glow of love beads and sandals. Needed was a "softer, more conformable product with improved PV"—water vapor permeability, or breathability—"and whatever else is needed to substantially improve foot comfort." And lower prices. And a more leatherlike feel.

Corfam had early on faced a veritable bazaar of synthetic competitors, with names like Arnav and Barretta, Morimer, Pervel, and Poron, drawn to the market first by Corfam's promise and then its early success. Anybody who'd ever made anything of plastic and could contrive a way to hype it as high tech, and maybe suggest breathability, real or bogus, had got onto the Corfam bandwagon. By fall 1964, just six months after Corfam's debut, enough of these materials were going into shoes that *Boot and Shoe Recorder* could herald "The Fall Flight of the Man-Mades" and tell of mock alligator pumps from Philips right alongside new offerings of Corfam. "Du Pont's success," it declared, "has encouraged many imitators."

But in the end it wasn't these Corfam wannabes that hurt it most but vinyl, its tawdry down-market competitor, unaccountably making headway in shoe uppers, driving down prices. "We realize we're doing some ice breaking for other man-made materials," Charles Lynch was quoted as saying at the time of Corfam's introduction. "But the 'free ride' isn't going to be as smooth as some may believe." Certainly not, surely, for vinyl, which already had a niche to itself—at the bottom of the market, where people couldn't afford to be fussy about foot comfort, with its reputation for tackiness, poor quality, and sweaty feet. The newer synthetic uppers, whatever they were made of, mostly avoided references to vinyl or, really, any specific plastic. One was a "multi-layer fused polymeric," another a "chemically expanded elastomeric," a third simply a "synthetic resinous material." Du Pont, of course, had from the start done all it could to distance Corfam from vinyl, aiming it at the top of the market, refusing even to call it a plastic.

But Corfam's own initial allure benefited the cheaper synthetics. As Bill Lawson put it later, "The advent of a poromeric material as an acceptable quality replacement for leather in shoes apparently lent respectability to the market efforts of the vinyl-coated fabrics producers." In other words, if Corfam was supposed to be so good, maybe just being a synthetic wasn't so bad after all. Footwear's simple class structure—up-market leather, bargain-basement vinyl—was now confused. Corfam? Vinyl? Poromerics? Polymers? "The customer doesn't know the difference," said a *Footwear News* article in 1971, describing the view of many shoe retailers and manufacturers.

But Corfam was far costlier than vinyl. In its first year it sold for about $1.10 per square foot. That figure is easy to misread today as maybe not *really* so much but, believe me, it was. Cheap shoes back then could be had for $3.98 a pair, high-end shoes for $20; gasoline went for 29 cents a gallon and a stamp cost a nickel. In 2005 dollars, Corfam was selling for close to $7 a square foot; or, in terms of fabrics sold by the running yard, 60 to 100 bucks a yard. Leather in the mid-1960s, meanwhile, went for 50 or 60 cents a square foot, many man-made upper materials for 25 or 30 cents. Even after modest price cuts, Corfam was *very* expensive. Sixty-five cents a foot might have given Corfam a shot in women's shoes, a Du Pont marketing memo estimated in 1969, but they never got close. "We want a meaningful reduction in Corfam's price," a Du Pont delegation told the International Shoe Company, one of the larger shoe manufacturers, in 1964. But first, it conceded, Old Hickory had to be made to run efficiently. "Du Pont's finest engineers are confident it will." As we've seen, it never did.

Du Pont had done all it could to keep Corfam away from vinyl, yet now the two were swimming together in the same fetid pool, struggling for market share. "We didn't think vinyl-coated fabrics would be tolerated by the general public," says Richard E. Heckert, who in 1969 had taken over as general manager of F&F, "but boy, were we wrong." By 1971 vinyl had one-third of the footwear market; all the poromerics together, Corfam included, less than 5 percent.

In late 1964 the Old Hickory plant went online and Du Pont delivered about 3 million square feet of Corfam. In 1965, 10 million; in 1966, 20 million. In 1967, Corfam was introduced for products other than shoes, including luggage, handbags, watch straps, belts, and golf bags. This was progress, was it not? And yet, as he looked back later, Bill Lawson would see the year 1967 as a turning point. From the shoe stores—which were on the front line, seeing their customers daily, fitting them, hearing their complaints—the reports of comfort and fit problems multiplied. It was not just waning enthusiasm that Corfam increasingly faced but outright hostility. A Chicago regional sales report in 1970 noted "continued retail opposition toward Corfam. [One shoe manufacturer's] salesmen have been 'turned-off' somewhat on Corfam because of the constant negative attitude of their customers. I have been told that many customers completely tune out whenever Corfam is mentioned."

The dark undertones of the dirge deepened. As early as September 1969, Du Pont was countering stories reported in *Footwear News* that it had "decided to get out of the poromeric upper business." Not so, it said, Corfam was doing fine, the company continuing to pump money into it. "We are confident of the future of this venture." A couple of months later, a troubling article about Du Pont came out in *Forbes*, highlighting Corfam's problems. "It was to do to leather goods what nylon did to silk stockings: Bury them." But it hadn't done so. Corfam was in trouble. "Time is running out," Heckert was quoted as saying. Oh, there was still hope. Make Corfam cheaper; huge markets would open up to it. "But as he speaks," the *Forbes* writer said of Heckert, "there are no more triumphal choruses and trumpet fanfares for Corfam." The story was such a downer that Joe Leavy, technical sales head, made up a list of five rebuttals to its doom-and-gloom message. "You obviously must understand," the Corfam loyalist was prompted to insist, "that the future for the shoe business is in man-made products." Why, the U.S. Department of Agriculture had recently predicted as much; the future was synthetic. "We are not going out of this shoe business."

But Du Pont engineer John Noble figured it was all over when word came down that the company was shutting down Old Hickory's Poromerics Research Laboratory. For John Korenko, a company auditor, it was the contrast from just two years earlier that stuck with him. Old Hickory in 1968 he'd found upbeat and forward-looking; but in 1970, "things were unraveling," people no longer so forthright, grimly deceiving themselves that things weren't so bad as they were.

In early March 1971, Du Pont's annual report noted the new, improved Corfam, introduced the previous year, that was supposed to have a softer hand than the original. But then, just eight days later, a press release went out from Du Pont corporate headquarters:

> WILMINGTON, Del., March 16—The Du Pont Company announced today that it plans to discontinue the manufacture and sale of "Corfam" poromeric material within 12 months.

Du Pont hands tell you of the magical ambience that ran through Old Hickory in the Corfam years, the sense that they were embarked on making a worthy and wonderful new material. From late 1964, when the plant opened, to 1969 were the golden years. "The Corfam group had togetherness in work and play I never again experienced in my 28 years with Du Pont," James Noble told me. Another remembered it as the closest, tightest group he'd ever been part of. It is the sort of recollection, with memories of youthful camaraderie, golf and tennis together, Friday night dances, the spark rekindled by later reunions, that more normally surrounds the memory of some great success; this time it applied to a failure.

How much did Du Pont lose? According to *U.S. News and World Report*, $250 million; by another account a cool billion, in today's dollars. Du Pont's own figure, in 1971 dollars, was $80 million to $100 million. Whatever the numbers, Corfam could never be written off as just another marketing mishap. It had benefited from too much technical expertise, been introduced with too much fanfare, and been promoted too confidently to just quietly disappear. Back in the 1950s *Du Pont Magazine* had run an article explaining just how and when the company withdrew products from the marketplace. Du Pont, said its then president, was like an open barrel with a spigot at the bottom: Research created new products, which went in at the top, while obsolete ones drained from the bottom: 120 products pulled in the previous four years, most soon forgotten, the process cool, considered, and under control. Corfam, however, wasn't like that; the barrel had abruptly ruptured, dumping its contents across the floor.

Corfam's fall inspired moral chastisement; some headlines verged on contemptuous or cruel. The *New York Times* described it as "Du Pont's $100 Million Edsel," referring to another colossal failure, a Ford car. The *Wall Street Journal* intimated that Corfam had been ignominiously fired—"given its final walking papers." A British newspaper presumed to tell its readers "How Du Pont's Corfam Took a $100m Tanning." *National Lampoon* imagined an Episcopalian Hell where all shoes were made of Corfam.

To recoup some of its losses, Du Pont sold its production equipment to a state concern in then-Communist Poland, Polimex-Cekop. The way the story's told by Ron Moltenbrey, the Du Pont engineer who coordinated the move and over the years made countless trips to Poland to help its new buyers set up, the Old Hickory machinery was loaded onto barges, floated down the Tennessee and Mississippi rivers to New Orleans, and then transferred to a Polish freighter. In Pionki, Poland, it was reassembled. By April 1975 the factory was up and running, churning out what Moltenbrey calls "military brown" for holsters, boots, and briefcases. Of course, the sale to Poland inspired its own memorable headline in *Forbes*:

HUNDRED MILLION DOLLAR POLISH JOKE

By July 1971, just four months after the announcement from Wilmington, a student at the University of Texas at Austin already had an M.B.A. thesis under his belt. It was called "The Corfam Failure." Corfam became a symbol of business failure; whenever discussion turned to marketplace disasters, the brain's synaptic pathways seemed to lead straight to it. It was included on a website devoted to "titanic flops," along with the Arch Deluxe hamburger and Premier smokeless cigarettes. In a book, alongside Edsel, New Coke, and Sony Betamax, as one of the 100 biggest branding failures of all time. In a speech, for the Center for Entrepreneurial Management, devoted to *Five Marketing Blunders*. When, at a New York celebration of successful advertising icons—Mr. Clean, Tony the Tiger, and so on—the discussion turned to failures, Corfam came up once again. In retrospect, it was easy enough to deride poor Du Pont and *fun, too*: "Women wanted breathable, comfortable shoes that would last until the next fashion trend, but Du Pont," the way someone put it 30 years later, "gave them hot, stiff uppers that lasted forever."

Barron's, however, saw in Corfam's collapse a different moral; it furnished proof that huge corporations were not immune to the risks of the marketplace, as some left-leaning economists said. If Du Pont, with all its marketing muscle and seemingly infinite resources, could fail, then presumably the market really was as treacherous as dyed-in-the-wool champions of laissez-faire had claimed all along. "Competition, like chemistry," the author concluded, "yields better things for better living."

Apparently responding to a stream of inquiries, Bill Lawson in 1972 wrote a frank and reflective review of the Corfam story. One section was entitled "Commercial Maturity and the Beginning of Doubt." The reasons for its failure were "subtle and complex," he wrote, and he offered a number of them. But in a departure from the buoyant self-assurance he'd displayed at the New York Hilton a decade earlier, when he'd compared the Corfam quest to *A Pilgrim's Progress*, a chastened Lawson suggested that readers feel free to draw their own conclusions: "There is still considerable speculation even among those who were intimately involved with the venture" about just what went so wrong.

It was surely time for a hint of doubt to wash over the Corfam project. Du Pont had been overly "dazzled by the glittering commercial prospects," as a post-shutdown *Fortune* article put it. Management had at first been insufficiently skeptical, then "slow to recognize the extent to which plans had gone awry." Caught up, in other words, in wishful thinking, a kind of self-delusion. In one narrow sense, then, there were not many reasons for the debacle, just one—the company's blindness to the plain facts of Corfam's weaknesses.

Whatever else it was as a slice of business history, Corfam was a reminder of the continued resilience and appeal of natural materials in an unnatural world and of the limitations of plastics that had, until then, seemed an implacable force of the future. In an essay for the *San Jose Mercury News* in 2003, Geoffrey Nunberg, of the Stanford Humanities Center, recounted the history of the word "plastics." For 40 years, he observed, the word had enjoyed a love affair with Americans. A 1940 poll found that "cellophane" was the third most beautiful word in the English language (behind "mother" and "memory"). Du Pont's pavilion at the 1939 World's Fair "featured a shapely Miss Chemistry reclining on a podium in nylon stockings." Nylon and Orlon, then Lucite, Formica, Saran, and the rest. . . . What Nunberg, a linguist, called "the break in the filament," however, came in the 1960s and was marked by two key events. One was "the first use of the word 'plastic' to refer to something superficial or insincere." The other, by the chemical giant that had given nylon to the world, was the introduction of Corfam and its subsequent market collapse.

In 1994, *Boston Globe* reporter Paul Hemp was asked to try, in the service of journalism, a new no-iron cotton shirt. "My boss thought I'd be the perfect guinea pig," he wrote, so he tried it. But the presumably wrinkle-resistant fabric, said to be imbued with molecular memory, did show wrinkles. And it exhibited a "polyesterlike, faintly metallic sheen." Hemp preferred his old rumpled cotton, saw the new shirts as just one more marketing dud. "Like the Corfam shoe and disposable paper dress," he concluded, "they may turn out to be a good idea whose time never quite came." The ghost of Corfam seemed able to leave on anything too unnatural a cold, repellent taint. A military website talked about the clip-on ties that soldiers sometimes used to get through dress inspections. "Clip-on ties," someone sneered, "fall into the same category as do those 'Corfam' patent leather shoes."

A writer for the *Dallas Morning News* wrote of his frustration with a prematurely fraying blanket, "one of those poly-something jobs without one whit of cotton, wool or anything else we hold dear." Pretty soon, he was extending his rant against all things synthetic to Corfam, recalling the "suffocating" feet of its victims. "I hope they buried Corfam with nuclear wastes somewhere in Idaho."

8
Top Grain

A nondescript trade show hall in the exurbs beyond Boston's outer beltway. Shoe retailers and shoe sales reps gather here for their twice-yearly show, talking up men's tasseled slip-ons, stilettos and Mary Janes, chukkas, sandals, and funny-looking footwear only a clown could love. Volleyball shoes, orthopedic shoes, brogues, shoes fit for the dance floor or the nursing ward. Shoes stacked neatly on shelves, arranged clumsily on tables, piled in heaps on the floor. Shoes everywhere. And none are made of Corfam. Some older sales reps remember Corfam from their early years in the business but many have never heard of it. Today, cheaper shoes may have uppers made of vinyl or other man-made materials. And Franco Sarto has a slinky, calf-clinging women's boot in black polyurethane that's gotten buzz for being expensive, sexy, and synthetic—an uncommon combination. Hi-tech athletic shoes, for hiking, or running, or wrestling, use synthetic materials, too, just as the classic sneaker of half a century before did. But mostly, in men's shoes or women's, whether waterproofed, napped, shiny, glossy, matte, or suede, whether in drab brown, jet black, or iridescent purple, the predominant material here, 40 years after Corfam, is leather.

Back in 1964, *Fortune*'s Lawrence Lessing concluded a rhapsodic paean to Corfam by asserting:

> The word "'synthetic" has lost most of its once mean, pejorative, nineteenth-century connotations. Instead, it now represents one of the major hopes of the world for adequately and decently clothing its explosively expanding population. Today the word "synthetic" refers to man's ability, through science and chemistry, to go beyond brute nature and create something new and better in the world—in this case, new shoes to walk in.

Maybe so, but Corfam failed to prove it. Lessing's "new shoes to walk in" were surely new, but they weren't better. Emerging as Top Grain, from this Seven Years' War of the shoe materials, was leather, in all its "brute nature." Its shortcomings were as plain as the tick bites that marred its grain or the frustration forever boiling up from cutters trying to navigate around them. Many of its virtues, on the other hand, resisted lucid description, much less numbers and graphs. But it had proved a worthier competitor by far than the men and women of Du Pont could have imagined. What *was* it about leather, anyway?

Part of the answer, ironically, was that the same piecewise and irregular nature that made for manufacturing headaches served it well in a marketplace dominated by a ceaseless quest for novelty and freshness. When fashion editors at the *New York Times* compared Corfam shoes to leather in an October 1964 article, one woman tester, while otherwise satisfied with her Corfam shoes, said she preferred "the variety of color, texture, and styling" she could get in leather shoes from high-fashion designers.

Were there, then, colors, textures, and styles that Corfam couldn't match? Hadn't Du Pont all along promised a wide variety of grain and surface "effects" for their new material? "Since the material is man-made," they'd said, "the appearance or aesthetic possibilities are limited only by the designers' imaginations." Corfam shoes might initially look like other footwear. But in time, "exciting new, non-derivative effects will be presented in high-fashion footwear, season after season—in appearance unlike anything ever seen before." Corfam could be printed, embossed, marbleized, metalized. There was nothing Du Pont's wizards couldn't do.

Couldn't do, that is, with enough time. But fashion was a volatile business, hectic and unpredictable. Each year, by one late 1960s estimate, 25,000 new or modified shoe patterns went into production, each inspiring choices in color and material; Du Pont executives admitted in 1969 they were surprised by how fast shoe styles changed. The big factory in Old Hickory that spluttered along making even plain-jane Corfam was like the proverbial supertanker that needs miles of open ocean in which to turn around; making even simple changes was a big deal. Corfam seemed forever behind. "With the fast-moving style changes in the fashion market," one shoe company executive was quoted as saying after Du Pont gave up on Corfam, "the emphasis is on speed." And they could be quicker with leather.

In principle, Corfam boasted all the advantages of consistency and repeatability that went with making any standardized product in a big factory (though, as we've seen, attaining it at Old Hickory had not proven easy). Leather, on the other hand, wasn't made that way in the first place, but hide by hide, skin by skin, using practices tailored, as they'd been for millennia, to the size and scale of the individual hide. "The thing that did us in," Joe Rivers recalled later, was that "leather comes in 8- to 10-foot hides. The colors and textures they could get were just amazing." At least when things were humming along at Old Hickory, Du Pont made thousands of yards of Corfam at a time, while "the leather people were inured to smaller batches" whose qualities, styles, and textures they could change on the run, offering special this or special that in response to the market. "It drove the Fabrics and Finishes people wild." What was most leather used for? Shoes, clothing, handbags, and gloves; people were fussy about the things they touched and wore on their bodies and sought delights to hand and eye that changed with the seasons. *Fashion.* Fashion meant constant turnover in the kinds of leathers used; for tanners, adaptability was a requirement for doing business. They had to respond instantly, could, and did.

Things have changed since 1972, but the Irving Tanning Company of Hartland, Maine, was in 2003 still responding to the twists and turns of fashion. Each season the company issues a glossy catalog—boats and buoys from the coast of Maine are this year's theme—showcasing its leathers, each with a catalog number and a name.

NIMBUS is so soft and fluffy it seems to float.

ESQUIRE, with a featureless matte finish, is so tight and hard that, especially in black, it's almost scary.

A leather called MUSKRAT is oily, with a soft, preworn look.

ANGELO is vegetable tanned, with a pebbly grain.

JAEGERMEISTER is destined for work boots, its waxy flesh side facing out.

PAZZO has a vaguely antique look, with a new grain laid down on top of the natural grain.

HEAVY METAL shows off a peculiar metallic luster when scrubbed or brushed.

STARBURST is shiny and showy, MIA bears a frosted surface glaze, BULLSEYE looks stately and expensive, like a Mercedes-Benz.

Each is not just a little different but dramatically so—as different one from another as any one of them is from Corfam, say, or Naugahyde; different enough, certainly, that you might admire one, respond to another with indifference, scorn a third. Packed into a colorfully printed heavy paperboard box the size of an unabridged dictionary are leather swatches of each, some in just three or four colors, others in more than 20: "Beach Grass," "Summer Taupe," "Gum Pink." Each swatch—there are hundreds of them—represents another fashion option, a distinct look, for work boots, dress shoes, belts, or bomber jackets. Each can be made as thin as a credit card or as beefy as a man's belt. And each is the product of its own custom-tailored manufacturing process.

One late winter day an order comes in from a shoe manufacturer for a waterproof leather the company calls RAPIDS, pitched as "a waxy full-grain tumbled leather. Handcrafted to provide rugged beauty much like that of Maine's coastline." The order is for 245 sides—a side is half a cowhide, cut along its backbone—enough for maybe 3,000 pairs of shoes, in a color called Yankee Barn, a chocolaty brown. On the tannery's routing slip an "F" is appended to the color designation, meaning the dye must reach all through the leather, worth an extra few cents per square foot on the final price.

One by one, a man passes the hides, still wet and blue from chrome tanning, through a machine that splits them to an approximate thickness; the top-grain layer is what the tannery processes, the inferior splits being sold for whatever they can get for them. Another man runs hides through a machine that shaves them down to their specified thickness, 2.2 millimeters, a fine powdery mass of blue leather shavings left deposited at its base.

Then, borne by elevator to a higher perch, all 245 hides are dumped into Wheel 6, a drum of African mahogany about as big, in three dimensions, as a small bedroom, its barrel-like staves held together with steel hoops, its interior all wooden pegs and shoulders that together work like an agitator on a home washing machine. Into it, over a span of perhaps 12 hours, are pumped dyes, oils, acids, and washes, in a programmed sequence of fillings-up and emptyings. All is monitored from a control room, filled with computer screens, set high over the great hall of whirling drums.

In succeeding operations the hides are "set out," vacuum dried, "staked," "milled." In setting out, rollers squeegee the wet hides. In staking—the name taken from its traditional counterpart when workers pulled and pushed the hides over the rounded ends of waist-high stakes set in the ground—the hides are mechanically pummeled, softening them. In milling, another softening operation, they're tumbled in large drums.

Before milling, these hides go to operation 1120, Hot Wax, in which melted wax is applied to the surface of the leather, its color a shade darker than that of the dye the hides already hold, so as to give them a "pull-up," or two-tone, effect when squeezed or stretched. Finally, about three weeks after their arrival, 239 sides, comprising 5,313 square feet of leather, worth about $15,000, are shipped out. Along the way, samples have been flex tested, color checked, and their moisture content and other variables regularly monitored.

All this comes after the tanning itself; most of Irving's hides, recall, come straight from the packer, already chrome tanned, in great, still-moist blue-green piles loaded onto pallets. True, tanning is the essential step in making leather. But here as in many modern tanneries the generic wet blues it's supplied are a starting point, not an end point. It is the tricky post-tanning steps that make some hides fit for shoe uppers and others for bomber jackets, leave some NIMBUS-light, others MUSKRAT-oily, and that today constitute the real art in leathermaking.

Indeed, these post-tanning "steps" *aren't* steps, not if by that we mean set, standard operations proceeding toward a particular end in a fixed order. For each leather demands quite different operations, sometimes in different orders, or maybe just tweaked a bit. In milling, for example, the leather is thrown into a shelved drum and tumbled, softened by sheer mechanical action; but depending on the leather you want, you might do it for 12 hours, half an hour, or not at all. You can dye leather, letting it soak up color in the drum. Or you can coat it with colored pigment, the hide passing along moving belts, under fast-spinning spray guns. Or both. And you can change your mind, be adaptable or finicky: The hides are not so soft as you'd like? Well, stake them again; pass them through the machine that peens them with vibrating metal fingers, softening them a bit more. Each of these and countless other options make for stunningly different leathers. So, of course, does the type of animal hide and how it's tanned. All together, they give the tanner the sort of fingertip control and flexibility that Corfam, in 1971, could only envy.

Of course, it was Corfam itself that helped galvanize the tanners into action. "Tanneries reacted to the introduction of Corfam with their most creative period of leather development," a story in *Footwear News* noted right after Du Pont pulled the plug. A trade journal in 1965 reported how the Irving Tanning of that era had a leather that resisted scuffing "as well as the synthetics." Today, four decades later, Irving and its global competitors, heirs to an ancient, almost insouciantly low-tech industry—*animal hides,* for goodness's sake—turn to ever more sophisticated chemical soups. British leather consultant Mike Redwood recites a litany of leather-waterproofing methods used since 1970: "Stuffing greases, hydrophobic oils, fluorocarbons, silicone treatments, chrome stearate compounds, dicarboxylic acid chains, and selected acrylic resins." American leather chemist Waldo Kallenberger points to new combination tannages, synthetic retannages, brighter, more stable colors, as well as the new waterproofing treatments, none of them around in the early Corfam days. A threatening new technology had kick-started an old established one; this, wrote London *Sunday Times* business writer Andrew Robertson at the time of Corfam's collapse, was the Sailing Ship Effect in action: "The best sailing ships were made after the coming of the steamer. If Corfam did nothing else it pushed the tanners into the late 20th century."

In 1964, reflecting on the challenge of synthetics at the 48th annual meeting of the Tanners Council in Chicago, Joseph C. Kaltenbacher, president of one of the big tanneries, stressed the need for tanners to "make a variety of different products—colors, finishes, textures, thicknesses, and feels—tailormade to our customers' changing needs. . . . This type of service," he predicted, "will not be easy for the machines of the imitators to copy."

And it wasn't.

Inevitably, Corfam focused new attention on the material it sought to displace. Once it hit the market, newspapers, consumer magazines, technical and trade journals judged not only its foot comfort, aesthetic qualities, and durability but leather's; it was as if a spotlight, first firmly trained on the synthetic, had jiggled loose and swung around to illuminate its natural rival. So by the time Du Pont gave up on Corfam, leather had probably been looked at more closely, more scientifically, more aesthetically, more every other way, than in all its long history—and often in a new, more flattering light.

For one thing, leather's intricate internal structure hove into view as a singular work of physical and biological beauty. Du Pont had tried to replicate it. But to state the obvious, Corfam wasn't leather, not even, strictly speaking, "artificial leather," in some ways not even close. As one researcher noted in 1968, leather's weave of fibers was much finer than that of poromerics, its mechanical integrity "achieved in more subtle fashion." Leather was mechanically superior, more resistant to fatigue. "There is little truth in the claim that the structure of Corfam closely reproduces the fibre weave of leather," a meeting of the International Union of Leather Chemists Societies in Lyon, France, was told in 1965. Corfam's "fibers"? They were "stout compared with the fibrils of leather" and, while interwoven, not nearly to the same extent. Corfam had a layered structure; leather's fibers ran up, down, and through, making for a single unified mass. "The structure of Corfam is similar to that of leather in the sense that a golf ball is similar to a football," which is to say scarcely at all.

Then, of course, there was the sheer comfort of leather shoes, long taken for granted, now reaffirmed. "Considerable discussion has taken place within the footwear industry regarding the inherent comfort properties of natural leather," noted British researcher R. E. Whitaker in a 1972 paper appearing in the *Journal of Coated Fibrous Materials*. The statement couldn't have been made 10 years earlier, before Corfam, which had sent industry scientists back to their labs to study the subject. They'd weighed socks and shoes, before and after wear. They'd asked shoe wearers about whether they experienced discomfort, and how, and at what time of day. They'd thought hard about dampness and tightness and just what they meant. Journal articles addressed foot health, shoe fit, the "microclimate" within shoes and carried titles like "Foot Comfort Properties of Natural and Artificial Leathers."

Corfam had been developed and sold as an upper material that breathed, that could shed foot perspiration. But there was more to foot comfort than that. For example, while Corfam did transmit water vapor through its microporous skin, any it failed to shed was apt to collect, leaving the foot in a soggy pool. Corfam absorbed little water; leather, by one measure, 13 times as much. When Corfam shoes were tested on school teachers, who were on their feet all day, the victims noticed; in one test, half complained of dampness—versus just one in five for leather. "It has been conclusively established," wrote L. G. Hole, head of materials research at Britain's Shoe and Allied Trades Research Association in a 1970 review, "that a leather upper provides a cooler and less sweaty shoe"—and, overall, a more comfortable one.

Put a Corfam shoe on a foot it doesn't perfectly fit and it stretches to accommodate bunion, big toe, or other fleshly protuberance, the foot made all too aware of the unwelcome pressure. At night the shoe reverts to the way it was. So next day, however uncomfortable it was the day before, it is still. Not so with leather. Corfam's problems could be defeated "by careful work at the fitting stool," Bill Lawson wrote in 1972. Leather shoes didn't need such careful work; up to a point, they adjusted on their own.

In 1963 an engineer described the following experiment: You arrange a round-headed plunger to depress a leather diaphragm, clamped around its periphery like a drum head, a set amount, three-sixteenths of an inch; it takes about a 30-pound force to do this. Then you withdraw the plunger. This pattern of mild abuse—PUSH DOWN and RELAX—is repeated 5,000 times. At the end of the test, which is supposed to mimic a 5-mile walk, the leather takes a permanent set; you see in it a cuplike protrusion corresponding to the plunger and reminiscent of the cheerful lumpiness of a worn pair of shoes. By that five-thousandth cycle—and here's the key point—the force needed to depress the leather that last little bit, to the full three-sixteenths of an inch mark, has dropped from 30 pounds to 7.

What the numbers explain is a long-familiar feature of human experience—of new shoes going on to feet they almost, but don't quite, fit; of initial irritation as the leather resists; of resistance weakening as the leather conforms; of the shoes coming to fit and finally holding their fit. "It is not known," said Mieth Maeser, the engineer describing this experiment to colleagues at an American Leather Chemists Association conference, "whether this property can be or has been achieved in the new synthetic upper materials which are being developed."

He meant Corfam, and it had not been.

Maeser, 63, had been named that year's John Arthur Wilson lecturer, among the association's highest honors. Three decades before, he'd come out of Utah with a bachelor's degree from Brigham Young University, gone east, and earned graduate degrees from MIT. There in New England, the heart of the American shoe industry, he'd taken a job in the research division of United Shoe Machinery Company, which furnished many of the specialized machines used to make shoes. That was 1934. Maeser had been there ever since, becoming an authority on properties of leather germane to shoe manufacturing. He was tall, thin, and uncommonly likeable, and everybody called him "Slim."

"Materials engineers," said Slim Maeser, "know less about our oldest, most traditional, and perhaps most used natural materials, such as stone, wood, and leather, than they do about any of our newer manufactured ones." Know less in part because such materials are so irregular and capricious, making it hard to define or predict their behavior. Over the years the technical societies had devised means to measure in leather what they could, quantify its variability. But large gaps remained and the material remained mysterious. Many were convinced—Maeser left no doubt he was among them—that the data "do not fully measure the true quality of leather. It may be," he added, "that leather has an elusive quality which is almost, if not entirely, subjective, and perhaps this quality is immeasurable because it means different things to different people, and yet is real to everyone."

Hard to measure yet real. *Slim Maeser was an engineer?!?*

Doing justice to his subject—"An Engineer Looks at Leather"—Maeser took his listeners on a tour of an imaginary shoe factory, reminded them of the stern rigors of lasting, noted leather's water vapor permeability and tensile strength, and, as we've seen, walked them through the leather diaphragm experiment. He spoke of how leather could be skived and scarfed without losing strength, how it held stitches close to an edge, how it resisted tearing at notches and stitch holes. But now, near the end of his talk, summing up, the engineer in him fell away and he showed another side of himself. Was it something like love for this material so long and so deeply part of his life? Leather, he said, was "an excellent and unique material." It had a "rich, fine, porous grain." It offered "beauty and texture."

And these, said Slim Maeser, the man from Utah, the MIT engineer, "cause a person to enjoy stroking it with his hand or rubbing it against his face."

Leather felt good, and this was central to its appeal.

The Du Pont engineers who developed Corfam were not indifferent to matters of texture and touch. But in the end, Corfam's perceived lack of softness contributed to its undoing. Just before the announcement of its demise, a *Fortune* article, entitled "100 Million Dollar Object Lesson," recounted its "aesthetic problems." "Shoemakers and purchasers alike were dissatisfied with what in a fabric would be called 'hand'—the feel of the material and the way it bends and creases and flexes in use." Dissatisfied, of course, compared to leather.

What the tanner worries about, declares a promotional booklet long used by the leather industry,

> is *feel*. He is ever mindful of his obligation to preserve in leather those characteristics which arouse the sense of beauty, and thus to produce leather having a natural feel. To fall short of this goal is to produce something with a synthetic hand to it—lacking the fine, luxurious feel of quality leather. It is of course true that imitation products can be printed, prodded, and patterned until some almost look like real leather. But the woman who for years has handled a beautifully aged crocodile leather handbag, or the man who has traveled the world over with an attaché case made of magnificent full grain cowhide, will quickly testify to the fact that products fashioned from leather have no equal.

In 1968, Du Pont had tests conducted for it by a Philadelphia market research organization, Associates for Research in Behavior. Fifty women were asked how they felt about a group of black dress pumps, identical save for the materials from which their uppers were made. While the women could not consistently identify the natural leather, and surely couldn't tell calf, say, from kid, they did prefer, in the words of the paper summarizing the results, "the visual and tactile aesthetics that are found in the best natural materials. . . . Women prefer a 'soft' leatherlike look to a 'stiff,' hard one."

Tanners and leather aficionados were, of course, forever harping on the lush softness of leather at its best. But what, exactly, was "soft"? The word itself, after all, can refer to the earth, to air, to rock music, to a pillow, to a kiss. It's experienced every day, all through life, and everyone presumably knows what it is. "More sleek thy skin. . . . And softer to the touch, than down of swans," wrote Dryden. But *define* it? *Measure* it? Early in his work, as we saw in the last chapter, John Piccard developed a primitive softness gauge for his experimental materials. In the years before and since, numerous other means have been devised with which to quantify it. Like the Tinius Olsen Stiffness Tester, and the BLC Softness Gauge, and the Shirley Softness Tester. In a method lending itself to easy, back-of-the-shop measurements, you need merely clamp one end of a leather sample to a table, let it drape over the side, and measure the angle at which it flops over; that's the Peirce Flex Test. At the other extreme, a British group tore samples of leather, recorded the sounds it made, and correlated sound wave forms showing up on the oscilloscope screen with, among other qualities, the leather's softness.

Yet would a score on one of these tests really predict a consumer's delight, or disappointment, in running her hands over a pair of leather boots? In practice, one industry chemist observed in an online forum, softness "is usually defined by the testing methodology," which is normally content to produce a number that doesn't pretend to measure anything like sensual or aesthetic appeal. Besides, even leather experts routinely disagree on what feels soft. So one review of softness instrumentation reasonably enough asked, "If individual assessors cannot agree on the order of softness of a group of leathers, then how can we ask a machine to agree with all assessors?"

When they try to express leather's sensual virtues, tanners turn to words like "feel," "hand," or "handle," typically relying on the punctuational help of quotation marks or italics to suggest the difficulty of speaking more precisely of anything so inscrutable. Depending on your definition—and the word has inspired many—"handle" is the sum total of one's sensations in touching, flexing, grasping, smoothing, or otherwise handling a material; it may or may not include visual sensations. Step into this uncertain realm and before long you'll hear the word "organoleptic," referring to tactile properties like handle, softness, and feel and extending to such ineffable questions as what it means to feel greasy, draggy, sticky, or smooth.

Plainly, words are part of the problem. Beginning in the 1970s, the textile industry, faced with kindred difficulties, turned to "Fabric Objective Measurement," inspired by the work of S. Kawabata in Japan, which sought to tie "handle" to a fabric's measurable physical and mechanical properties. There are "so many terms in use in English alone," noted a 1996 review of the field, "that confusion is almost inevitable, even before any attempt is made at translation into other languages." Smooth, soft, firm, coarse, thick, heavy, warm, harsh, stiff, lively, crisp, papery, greasy, weavy, boardy, drapy. These are among terms used to describe fabrics in English. Japanese has its own, such as *koshi*, with its suggestion of stiff-and-springy. Or *fukurami*, with conjoined intimations of bulky, warm, and full. Complicating matters further, the word you pick is apt to depend on how hard or soft the skin of your hands is, how rough or smooth, how warm, cold, sweaty, or dry.

Touch is uniquely two way. So explains Mandayam Srinivasan, director of the Laboratory for Human and Machine Haptics at MIT, more familiarly known as The Touch Lab. View an object and you're not *doing* anything to it. Same goes for the sense of hearing. But touch something and, even as it reaches out to you with sensations, your fingertips massage it. Run your hands across a piece of calfskin and you stir vibrations that depend on its surface features and those of your fingertips. Four distinct classes of sensors in the skin respond, each to a different range of frequencies; mechanoreceptors, Srinivasan calls them, the tactile equivalent of the specialized light receptors in the retina, its rods and cones. One class, for example, is most responsive to vibrations near 300 Hertz, or cycles per second, another to those around 30. It's how these classes of mechanoreceptors respond together that makes for the pattern interpreted by the brain. "Patterns that distinguish a mango from an orange," says Srinivasan, "have certain sets of neural signals." Likewise those distinguishing cowhide from, say, Naugahyde.

Consider the basketball research conducted in the early 1980s by Sharon Mathes and Kay Flatten, of the Department of Physical Education and Leisure Studies at Iowa State University. Sports were being invaded by new technologies, like aluminum bats and oversize tennis rackets. Maybe it was time to subject one of these newfangled innovations to scientific scrutiny. Their choice? Basketballs made of synthetic leather.

In one test they dropped the balls from a standard height, 6 feet, noted that the leather ones bounced higher, and concluded they could therefore be dribbled more easily; score one for the real thing.

But Mathes and Flatten were particularly interested in whether people could tell the difference between the two balls in the first place. They enlisted the help of male and female students taking a phys ed class.

From a distance of 4 feet, subjects looked at the balls and pronounced them either synthetic leather or real.

They listened to an experimenter dribbling them.

They dribbled the balls themselves while wearing goggles that interfered with their vision.

And, blindfolded, they ran their hands over them.

The sound of dribbling, it turns out, didn't help them distinguish one from the other; they were wrong about as often as they were right.

Likewise with viewing them; the synthetic balls looked more real than the real thing, fooling the test subjects two-thirds of the time.

It was only once the students got their hands on them that they could tell synthetic balls from leather. Even madly dribbling, their hands in contact with the ball only intermittently, they could "read" the synthetic material for what it was, picking correctly about two-thirds of the time. And all doubt evaporated once they could freely touch each ball, feel its heft, hold and grasp it as much as they liked. They could distinguish one from the other 82 percent of the time.

The resulting paper, which appeared in *Perceptual and Motor Skills,* spanned just three pages—just a simple compare-and-contrast sprinkled with a little chi-squared statistical test. But it spoke volumes: *Looks like leather*, from 4 feet away, anyway, was easy to achieve; *feels like leather* was another thing entirely.

Long ago, the story goes, the 5th-century B.C. Greek sculptor Myron made a bronze cow that was so realistic it left neighboring bulls sexually aroused and angered herders whose flocks it refused to follow. But touching it instantly gave it away. "To one who only views the cow, it will seem that art has stolen nature's power," William R. Newman recorded the epigram in his book *Promethean Ambitions*, "but to the onlooker who actually touches the animal, 'nature remains nature.'"

Perhaps artificial leathers, like Myron's cow, could never truly steal nature's power?

In 1971 at the English seaside resort of Blackpool, Britain's Shoe and Allied Trades Research Association convened an international conference devoted to artificial leathers. Among those on hand was Daniel J. Troy, a Du Pont chemical engineer. In 1966, Troy had become a research fellow with Fabrics & Finishes, Corfam's home department, where he'd grown interested in the aesthetics of poromeric materials; he studied how they looked. In developing Corfam, he said at a conference session, the aim was to make it look like leather because, first, that's what shoe buyers expected, and second, "as an indicator of future performance," thereby trading on leather's reputation. Matching leather's color and gloss, he reported, was easy. But texture was trickier. Humans could pick up on the subtlest imaginable differences. So subtle, in fact, that they verged on nondiscernible. And to explain the seeming paradox, he asked his largely British audience to consider the work of one of their countrymen, David Pye.

Pye was an architect, industrial designer, and furniture maker, a long-time professor of furniture design at the Royal College of Art in London. A few years before the SATRA conference, he'd written a wonderful, idiosyncratic book that became something of a cult classic, *The Nature and Art of Workmanship*. Most famously, Pye discriminated between what he called the "workmanship of risk" and the "workmanship of certainty," the first loosely that of the true craftsperson, the second of the factory operative. But that was just one of countless keen insights. Another was that, in responding to sculpture, music, furniture, or any work of art or craft, we are alive to the slightest deviations and irregularities, the smallest scratches and imperfections, even those lying at the very limit of perception. We don't quite see them, or maybe we do, sort of, but not quite, yet they affect us. As he put it:

> In nature, and in all good design, the diversity in scale of the formal elements is such that at any range, in any light, some elements are on or very near the threshold of visibility. . . . The elements that at any given range, long or short, are just at the threshold, that we can just begin to read, though indistinctly, are of great importance aesthetically. They are perhaps analogous to the overtones of notes [in music]. They are a vitalizing element in the visible scene.

And they contribute to our pleasure and satisfaction. In leather it might be little hair cells, slight variations in pattern, blemishes too small for even the most careful cutter to exclude, inconspicuous variegations of shade, color, or texture. A tag accompanying a hat made by the Henschel Hat Company of St. Louis assures us that "the variations in color, scars and even insect bites are visual proof this item is made of genuine leather." The tag draws the consumer's attention to natural markings he can plainly see. But Pye was going further, suggesting that even those you don't *quite* see, or aren't conscious of seeing, count, too.

Pye's insight resonates with research described by Christopher Moore, a cognitive scientist at MIT's McGovern Institute for Brain Research. Excite a touch receptor in a human subject's skin with a "loud" stimulus, he explains, and she feels it, duly reporting it. Excite it with too soft a stimulus, and she reports nothing. This much is obvious. Now strengthen the stimulus to at, or just barely above, a level previously established as the threshold of feeling, and what happens? Sometimes, not much; not quite consciously aware of the ever-so-light pinprick, the subject volunteers nothing. But now, finally, in a key experimental variation, *demand* an answer with each trial: *Do you feel something or not—yes or no?* In this "forced-choice" situation the subject more consistently, and correctly, registers awareness of threshold-level stimulation; she "knows" without knowing.

One time my mother offered me a taste of a lasagna sauce she was making.

"Mmmm," I said, "It's great."

"Could you taste the coffee?"

Coffee?!?

No, I said, incredulous, I could not taste the coffee. I could taste that it was delicious, that *something* made it distinctive, but certainly nothing I could identify as coffee.

"I just gave it a spike," she said—a "spike," in a different sense realm from Pye's, that worked on me whether I was aware of it or not.

Pye also observed that, while natural materials have in profusion such barely perceivable yet aesthetically crucial features, man-made materials often do not. The markings or elements he had in mind are

> found in every natural scene except for those which most depress us, like a white wall of fog, or an evenly overcast white-grey sky. But they are not always found in the environment man has made for himself, though formerly they always were [as in things made in simpler times with simpler tools]. That explains the blankness we often find, now, in the expression of a product or a building when we get close to it. Down to a certain distance everything about them looks well. As you close the range after that point nothing new appears. There are no further [visual] incidents. As soon as you get towards the minimum effective range of the larger, designed elements, the whole thing goes empty.

Could Pye's insight help explain the weak visceral hold synthetic materials often have on consumers? It was one thing to look across a room and see real and faux leather shoes that looked about the same. But close up? Or in your hands, touching them? "The Corfam shoes looked like good leather," *Consumer Reports* concluded, but "they did not quite feel like it to the touch." Corfam was no Naugahyde, no slab of vinyl melted onto a piece of cloth; compared to earlier imitation leathers, it was a complex and subtle material. But compared to leather, and the mysterious, twisted history of life and growth embodied in it, with its tick bites, fat wrinkles, and flaws, all subject to the vagaries of liming, bating, tanning, fatliquoring, staking, milling, and the rest, it was simplicity itself. For Du Pont the exasperating irregularity of animal hides argued for Corfam, with its vaunted "uniformity in thickness, width, hand, color and texture." But from Pye's perspective, such bland homogeneity meant something was *missing* from it.

Daniel Troy, Du Pont's man at the Blackpool conference, must have realized Corfam had a problem. He recalled a Corfam style from 1966 designed to compete with black calfskin that had been "described by some observers as grey and sugary." Yet *that* Corfam, he said, was better than an earlier version he'd heard called 'hungry.'" Whatever leather is, it isn't hungry. "It commands sumptuous textures, which are improved by polishing and preservative agents," a booklet accompanying an early 1950s museum exhibition said of it. "Visually, leather is ornamental, tactually it is inviting; wherever luxurious surfaces, a smooth and stable brilliance, richness of lustre and depth of sheen are wanted, it has always been the optimum."

And not just when new.

An editor at *Leather and Shoes* described his first exposure to Corfam at a "bang-up coming out party" Du Pont held for it the previous week, where writers and journalists had gotten a chance "to feel, twist, poke, dampen and in general to get chummy with the new product." There might be a place for Corfam, but he suspected that men, at least, would not "settle for the product displayed at the Du Pont show." Corfam *did* "look new longer." But that was just the problem: A man "likes his shoes better when they have become somewhat aged, where quality leather takes on the breaks and flex marks indigenous to his own feet."

Here was part of leather's appeal: An article made from it was somehow never quite "new," or at least not *just* new; almost reflexively, you thought ahead to when it would be old and worn—and often better. Worn leather shoes were apt to fit better; we've seen that. But the same went for motorcycle jackets, handbags, and upholstery. Things made of leather become softer and warmer, their polished glow deeper. With synthetic materials, it could seem, they were never so good as when you brought them home from the store; after that it was all downhill. A vinyl-covered bicycle seat leaned against a brick wall a few too many times is left ragged and torn, exposing the yellow foam beneath, ruined. A leather seat suffering similar abuse shows the wear, certainly, but any abrasion, roughness, or discoloration typically seems more scratch than wound.

We ask of "quality" goods that they last a long time, not break down the moment they're put to heavy use. By this yardstick, leather goods often qualify—but *not*, normally, because they look new longer, rather because they grow old gracefully. As it's flung on tables, opened and closed repeatedly, and absorbs sweat, a leather notebook cover shows marks and irregularities that make it more appealing, not less. A wallet is burnished every time you pull it from your pocket. Leather upholstery gets softer and more inviting the more you plump down into it. All those lab tests for leather and faux leather—fading, crocking, cold-crack, stitch-tear, so many hours in the sun, so much rubbing and pummeling, so many thousands of flexes—say nothing of the material when new, but rather how it will hold up over time. The affection leather inspires owes much to its dignified aging.

In 2003, I attended a conference of the American Leather Chemists Association whose terrain ranged, across its three days, from the caked manure of the feedlot, to the tannery, to the testing lab, to the quiet interiors of plushly appointed luxury cars—a material world full and rich to the eye, to the mind, and maybe to the heart, too. "There is a love affair," said one speaker, "between leather and the human being."

"For some reason," Mieth Maeser said during his 1963 Wilson lecture,

> people seem to prefer the natural products for uses where their bodies come in intimate and continued contact with them. People like to feel a piece of polished stone but not a piece of plaster; they like wooden floors and wooden furniture; they like soft cotton and silk fabrics next to their bodies; and they react more favorably to leather bags and leather shoes than to similar products made of plastics.

Nancy Kanwisher, a cognitive scientist also at the McGovern Institute, has suggested that our brains have neural machinery to perceive human faces, a capability evolved over hundreds of thousands of years. Might we likewise respond with special pleasure or familiarity to the feel of wood or leather because of similarly acute neural sensitivity, tactile as well as visual, to materials with the deepest roots in human history?

In those buoyant years when Corfam was still a poster child for synthetics and anything a Du Pont man had to say about it carried almost oracular significance, Bill Lawson wrote a far-ranging essay for a trade magazine, introduced by its editor this way:

> Debate and discussion about the merits and future of man-made shoe upper materials has run the emotional gamut from condemnation to exaltation. What has long been needed is a deep, objective view. The article presented here is just such a view—and undoubtedly the clearest and fairest thinking ever published on this controversial subject.

Lawson, a chemical engineer by training, allowed that "when an animal synthesizes a polymer to serve as a component of its skin, it does an admirable job." But when humans remove the skin and use it to sheathe a human foot, say, they must alter that polymer; that is, they have to tan it, make it into leather, with all its glaring insufficiencies. Coming was a better day—Corfam was harbinger of it—when "the chemical industry can synthesize a polymer for almost any purpose and achieve a more satisfactory product" than any nature offered. The rest of Lawson's essay was given over to how chemists had systematically studied leather, designed around its shortcomings, and created something better. That was 1965. Six years later Corfam was dead, and leather was in its golden age.

Back in 1913 word circulated that the French army had ordered 200,000 pieces of military gear traditionally made of leather, only this time of a "specially worked cotton" perhaps like Fabrikoid. "Is leather's reign over?" asked René Madru, of Paris's Leather Experimental Station. "Will other ersatz materials, other pseudo-leathers more or less artificial, come to dethrone this ancient material that has always aided man in his work, granted him his triumphs, and never let him down?"

He didn't think so: It was one thing to take any one of leather's qualities and imitate it. But take from leather's whole repertoire of virtues, from its seemingly bottomless kitbag of distinctive traits? No, that was different. Leather, Madru affirmed, would always come out on top.

All during the 1960s Du Pont officials kept careful track of their many synthetic competitors, listing and describing the technical features and market prospects of each. One of them was Hitelac, which comprised a nonwoven substrate impregnated with nylon and polyester, a woven interlayer, and microporous coating. "Very poor aesthetics," a Du Pont memo said, "low moisture vapor permeability, poor break. We . . . do not consider it a real competitor now." Hitelac was manufactured by the Toyo Rayon company of Japan. In 1970, Toyo changed its name to Toray and that year introduced a synthetic suede that in the United States became known as Ultrasuede.

Another competitor was Clarino, which by 1967 had most of the Japanese market for poromeric shoe uppers. In America it relied little on the lavish advertising that had fueled Corfam's early growth, but even so earned substantial business from shoe manufacturers. Clarino was manufactured by the Kurashiki Rayon Co., of Osaka, Japan, which in 1970 became Kuraray Co., Ltd.

Part II

Inspired Fakes

9

What Nature Had in Mind

The glint in the distance, the burst of brightness against the dry scrubby plain, could be flocks of sheep or cactus leaves. The sheep are familiar enough puffs of white. But when clusters of native cacti, with leaves like flat plates reminiscent of Triceratops armor, catch the Mediterranean sun like mirrors, they, too, gleam white. We are in the valley of the river Tirso, near the middle of the Italian island of Sardinia. Hills rise above a landscape of stone outcroppings, mortarless stone walls, grazing sheep, cacti, bushes, low trees. But its most prominent feature, visible from a distance across the valley floor, is a pair of smokestacks that tower like exclamation points over a factory complex. Here, near the town of Ottana, in a setting of nondescript industrial structures that could have leapt intact from the American Midwest, they make Lorica. "*Como la pelle,*" Signor Savini tells me. Like leather.

Giorgio Savini wears a closely Euro-fitted gray suit over a trim body he holds straight and tall. He has a deep tan this March day and a neat black mustache; only deep pouches under his eyes mar his Marcel Mastroianni good looks. He is 59 now, the company's director of research and production development, working out of an office back on the mainland, in Torino, in the foothills of the Alps, home of the 2006 winter Olympics. But he's here today, as he often is, for sometimes two weeks a month, to confer with his colleagues Giuseppe Munafò, plant manager, and Enrico Racheli, assistant to the company's CEO, who is in for the day from company headquarters in Milan. "Beyond leather," is the company's own boast for Lorica. But endorsements come from outside, too. "The most amazing 'synthetic' leather ever developed," says Waldo Kallenberger, the Cincinnati chemist who's spent all his professional life in the hide and leather business. "This stuff could actually do what the Corfams and Naugahydes threatened to do to leather decades ago." And here in Sardinia is where they make it.

Or rather, *finish* making it; Lorica isn't made here from scratch. Arriving at this two-story factory employing about 40 people are mostly not raw fibers and chemicals but big rolls of "substrate," the material to which the company's technical legerdemain is applied. Savini and his colleagues call them greige goods, from the Italian *greggio*, meaning gray, an old textiles term for unbleached, undyed fabric. It resembles the final faux leather but is an off-white, with just a suggestion of grain, and feels rubbery. It's produced halfway around the world, by a company with which Lorica has dealt for more than two decades—the Kuraray Co., Ltd., of Osaka, Japan.

Savini hadn't heard much of Kuraray until one day in the fall of 1982. He was working for one of Italy's largest tanneries, Conceria Italiana Reunite, or CIR—pronounced CHEER—in Torino. "How will I survive?" he'd asked himself the first time he breathed the tannery's sulfurous air. But he found the tests of chrome tanning he did for the company more interesting than anything he'd studied at the university and soon was helping to develop new finishes and working with customers. That's how things stood when in 1982 he made the obligatory seasonal trip to Lineapelle.

Lineapelle is the great Italian leather fair, a three-day extravaganza of art and technology, color and glitz, that brings tanners together with fashion designers, upholsterers, and clothing and shoe manufacturers to show off new colors, new finishes, new looks. These days, it's held in Bologna; then it was in Milan. CIR's large booth had three meeting rooms. And in one of them, while Savini presided over the display, his boss, Giorgio Poletto, R&D director, met with a delegation from Kuraray, maker of Clarino, the artificial leather against which Corfam had competed in the 1960s.

Kuraray had a proposition for them. They'd gone as far as they could with faux leather, wanted to take it further, and were coming to CIR for help. "This," Poletto told Savini later, handing him some samples, "is their new synthetic leather. It looks terrible, I know. But it can be improved." Closing up on Friday night, Savini forgot all about it. But back in the lab on Monday, Poletto was right there with the Kuraray samples, four or five of them, each a foot and a half square, in various compositions and thicknesses. Kuraray was asking them, makers of real leather, to make synthetic leather better. And this was their starting point. Poletto was "a strong person, always looking for something new," says long-time friend Savini, speaking through Racheli as interpreter. Why *wouldn't* CIR be interested? Leather was slow just then. Maybe in whatever Kuraray brought with them from Japan was a way into new markets. It was worth a try.

Worth a try doing what exactly? They certainly weren't going to tan this polyurethane and nylon microfiber material; how could you tan something that couldn't be tanned, that didn't *need* tanning. But retanning? That was another story. In normal tannery practice, chrome-tanned wet blues scarcely resembled anything you'd call leather. Typically, they'd go through additional steps, today broadly called retanning—that soaked them in dyes, fats, and oils, and mechanically worked them, to improve their look and feel. Could they do something like that to the lifeless patches of plastic they'd gotten from Kuraray?

They placed samples in drums, then oiled and dyed and tumbled them. At first, they got nothing for their trouble but cracking and peeling, ugly spider web wrinkles. In coming months they asked Kuraray for more samples, with varying thicknesses of polyurethane film, some embossed with a heavy leather grain, some only lightly. In a corner of the tannery they set up a little experimental enclave, with miniature drums. They'd begin work late in the afternoon and carry on into the evening.

Needless to say Savini and Poletto kept their day jobs; on the strength of their initial failures, they were hardly ready to commit to this new venture. Around the tannery, Savini remembers, "They said we were crazy." Some were upset with them for messing with materials that competed with leather. Savini had his own doubts. He'd spent his working life with leather. And now he was carrying on with this synthetic interloper? Why, it was like an illicit affair—in the morning with your wife, in the afternoon your mistress. The sense of divided loyalties gnawed at him.

The real crisis came when they actually got something to work. Start with lightly embossed substrate, tumble it in milling drums, much as they would real leather, the light impress serving as nucleus or starter, and they'd wind up with a deeper, more attractive grain. "You give it a place to start, for the wrinkles and grain to develop," explains Savini. You can see it at work today at Ottana. After they slice the starter material to the proper thickness, then soak and dye it, its surface lacks definition, looks dead. But then they load it in milling drums like those used by tanneries all over the world, and 6 to 8 hours later, it comes out softer and warmer, with a pronounced and appealing grain, more like leather.

This modest success, though, only raised the stakes: What to do now? Poletto was 43, Savini four years younger. For him the tannery represented his professional coming of age. He'd made his name in the industry, had his roots there. Change directions now? What would the market say about their new *unleather*? With a harsh verdict, would the whole precarious enterprise abruptly keel over?

But in the end they were driven by curiosity, the sheer spirit of adventure. In Italy, leather had a nobility to it, true, but in many ways it was a stodgy old business. "From the R&D point of view," Savini says, "microfiber was more interesting. . . . It would be a challenge to give life to plastic. To give something natural to an artificial material." To take that cold dead substrate shipped in from Japan and twist it to their designs—the very prospect was exciting. They'd stay together, the two old friends, work with this strange material at the cusp between two manufacturing traditions, so faux and yet, through their work with it together, real to them. So first at CIR, then through various ownership changes, and on to the new company, Lorica, they kept at it.

The greige came in rolls. And the final product would go out in rolls. But they realized soon enough that they couldn't just unwind the stuff, dump it into drums built for treating hides, and expect it to come out right. A hide, as the logos of tanneries and leather trade organizations almost ritually show, has a shape to it, a definite proportion—a meter wide, say, by a meter and a half long. Long ribbons of greige, on the other hand—any much longer than double their width—would get all knotted and tangled in the drums. The solution, still used, had no high-tech pretensions: They fold up 15-meter lengths of material and sew them up into neat, contained, looping bundles just $2^1/_2$ meters long. Once finished sloshing around in the drums with oils, dyes, and proprietary chemical brews, the bundles are unloaded by a man in rubber apron and boots who looks as if he could have stepped from any tannery in Italy. At this point they feel like a big load of jeans and cotton shirts pulled from the clothes washer and bound for the dryer—heavy, clammy, and cold. With big shears, the aproned man unsnips the stitching and lays it out at full length on a stainless steel table. After drying and sewing up again, it's on to the milling drums to soften it and bring out its grain and then to spray-finishing for surface color.

Their Japanese clients were surprised by what they'd achieved, especially its softness, and saw the prospect of big new markets. Says Savini with some pride: "This was not their know-how"; it was *tanner's* know-how. His work with Poletto was embodied in several patents, one of which was U.S. Patent 4,766,014, "Process for Producing Artificial Leather Similar to Real Leather by Chemically Processing Synthetic Sheet Material," issued in 1988.

Designers with the Italian fashion house Trussardi came up with the name Lorica for the new material; loricas were light body armor worn by the ancient Romans. At its introduction at the Piazza Roma in Milan in 1987, Savini remembers, stairs illuminated by torches and lined with models of loricas led up to a platform on which stood sofa and chairs upholstered with it. A hundred journalists were on hand. Though Lorica had at least one fashion show later, in Nice, that big Milan splash was the high point; no Corfam-like promotional bludgeoning of the market. But 20 years later, Lorica's been around more than twice as long as Corfam ever was.

Savini and Poletto figured furniture to be Lorica's natural market. Their Japanese partner, Kuraray, thought shoes. "In Italy," says one of the Lorica troika, all wearing fine leather dress shoes, "we can't imagine a fashion shoe made from synthetic." But what the company today calls "technical and sporting footwear" was another story. On the second floor of the Ottana factory, beside the offices, a display case shows off products made of Lorica, most of them traditional leatherware—gloves, upholstery, handbags, watchbands. And cycling shoes; these and other specialized athletic shoes have proved a particularly fruitful niche. In America, Doc Martens chose Lorica for its "vegetarian" shoes, aimed at select boutiques in Boston, New York, and Los Angeles. Out of the Ottana factory, which started up in 1991, following an earlier period when Lorica was still made in the tannery, come 600,000 meters of Lorica per year; that's about as much as

Du Pont made of Corfam in its second year at Old Hickory. It comes as thin as four-tenths of a millimeter, as thick as 2 millimeters. Some of it's buffed into suede; much more of it is smooth-grained. "First we had only leather," declares the company. "Today we have Lorica."

Especially on the flesh side, where leather aficionados often look first, it's easy to mistake it for leather. Under a low glancing light, you see the same welcoming little tufts of fine fiber and feel the same resistance to the slip of fingertips across it. Its grain side is not quite so successful, affords less pleasure to the touch. And Lorica lacks leather's sheer heft, though that's sometimes touted as an advantage. But snip it with a knife, then try ripping it apart, and the tear won't readily spread; drawn, fibrous tufts gathered at the tear testify to the violence of the attempt. "It's great stuff," asserted one online correspondent. When a distributor of vegan condoms working out of West Salt Lake City, Utah, sought material with which to make vegan bondage gear a few years ago—all black, of course, and more customarily made of leather—he settled on Lorica. "It feels great and it looks like leather," a wire service report quoted a customer. Even well worn, Lorica can pass: In a conference room at the Ottana factory, chairs upholstered in Lorica show their age, but genteelly: Leaned against walls, rubbed up against desks, or otherwise abused, some of the tan or chocolate-brown seat backs are abraded by use. But no underlying fabric shows through, no peel, nothing is cracked—just that familiar peach fuzz of wear that could as well be leather.

"A revolutionary material made by an exclusive system for processing very fine microfibres." That's the company's own take on Lorica. But if truly revolutionary, how much was owed its finishing operations? And how much to the substrate, with its mysterious "microfibres," from Japan? Savini, Racheli, and Munafò keep in touch with their Japanese partners; they have visited Japan, seen needlepunching and other processes that, as in Corfam, go into the substrate. But *not* the ones that leave its maker, Kuraray, claiming to make a true "man-made leather," not just another sad "artificial leather" of generations past.

The faux leathers produced by Kuraray, Lorica, and many other companies are all based on microfibers. Over the past 40 years, at least, they have been almost entirely the work of the Japanese.

Leather has never been big in Japan, just as meat eating isn't even today. The country is small. Animals demand large grazing areas. The leather there was typically worked by the lowest elements of society, the Eta, or Burakumin. A much larger feature of Japanese material life, on the other hand, was paper; think origami, shoji screens, lacquered paper boxes. In the 1880s the Prussian ministry of commerce sent Professor Johann Justus Rein to Japan to study its arts and industries. The country's paper particularly impressed him. It went into familiar articles like books and wallpaper, he reported, but also window panes and handkerchiefs, umbrellas and parasols, string and cloth. And it was used in place of leather; Japan's earliest faux leather tradition goes back at least 300 years and is based on paper.

Beginning in the early 17th century, the Dutch introduced fabulous gilt leather, or Spanish leather, to Japan. Originating in Spain, it had long been used to cover chairs and sofas but also—gilded, dyed, tooled, and elaborately decorated—for wall panels in baronial dining halls and by feudal lords for saddles and scabbards. Soon, the ancient arts of long-fibered handmade Japanese paper, or *washi*, were put to work imitating it. Early examples of Japanese leather paper go back to the late 17th century, when it was used for tobacco pouches permitted in nature-worshipping Shinto shrines that barred articles of leather. The Japanese imitated "so admirably the stamped and embossed leather in use in Europe," reported Sir Rutherford Alcock, first British ambassador to Japan, "that when specimens first came under my notice I had difficulty in believing the imitation of the royal arms of Holland . . . was native work."

During the mid-1800s, before the opening to the West and the introduction of industrial paper technology, 100,000 households made *washi* from bark. "When I was young," a Western visitor to a Japanese village was told in 1983, "everyone around here made washi—60 or 70 houses all told. I remember when, on a sunny day, there were drying boards all over the hillsides." The long, labor-intensive process yielded paper softer, tougher, and more flexible than its Western counterparts. Paper destined to become leather paper went to a special press that left it lightly crinkled. It was then spread out on a board, smooth side up; coated with thin rice paste to which lampblack had been added; dried; paste dye applied to it; and finally impregnated with lacquer.

The likeness to leather was, of course, imperfect. Alcock complained that "the difficulty of getting rid of the smell of the oil or varnish used in the manufacture is . . . the chief obstacle to the substitution of this article in Europe for leather in book-binding, chair-covering" and other uses. So maybe you wouldn't *quite* mistake it for the real thing. But it was often used for just those articles you'd expect to see in leather, like letter portfolios and small boxes. Greg Herringshaw, a wallcoverings curator at the Cooper-Hewitt Museum in New York, showed me samples of Spanish leather still bearing tack holes where they'd been mounted to walls or furniture and, beside them, *kinkarakawa*, leather paper worked by modern craftsmen, boasting all the substance, relief, and design intricacy, all arabesques and geometric figures, of the original. Turn one over to its "flesh" side, and you won't be immediately sure just what it is, its play of entangled plant fibers reminiscent of leather's. Herr Rein, the Prussian professor, lauded Japanese leather paper for its "beautiful appearance, surprising elasticity, and a softness that reminds one of calf-leather."

The heyday of Japanese leather paper didn't extend much beyond the 40 years beginning about 1880. Quality fell. The fad passed. But the episode could be seen as an example of mutual imitation at work across years and around the world. Spanish leather, itself an imitation of fine textiles, was imported to Japan. There it was imitated in paper. Then, in an era when all things Japanese were popular, Japanese leather paper traveled to Europe, where it was seen as fresh and original. A century later a new generation of Japanese man-made leathers would make a deeper, more longer-lasting mark on the West.

Around the time of leather paper's decline in Europe, the teens and 20s of the 20th century, numerous chemical fiber companies were born in Japan; it was the first great growth spurt of synthetic fibers, in particular rayon. Teijin, formed in 1918, tried to build up its rayon business straight from the patent literature. Asahi Chemical, which started up after World War I, bought its rayon technology from a German company. Kuraray was Kurashiki Rayon Company and before that Kurashiki Kenshoku, its name when founded in 1926. Toyo Rayon, which became Toray, was formed the same year. Ultimately, these and a few others became giants of the Japanese chemical industry, like Dow or Du Pont in America or ICI in Britain. Today all produce synthetic leathers, such as Lorica and Ultrasuede, rooted in a new technology, microfibers. According to John E. Berkowitch, in a

1996 report for the U.S. Department of Commerce, microfibers marked nothing less than "a new era in the history of synthetics."

The Berkowitch report was called "Trends in Japanese Textile Technology"; indeed, we have arrived in another country now, that of textiles. For early faux leathers like Fabrikoid or Naugahyde, the fabric on which the coating was laid down, cotton or muslin, say, was almost incidental. For Corfam the ordinary polyester going into it, made in a factory next door at Old Hickory, was indistinguishable from that going into shirts and leisure suits by the thousands of tons. But now with these new faux leathers, it was the textiles themselves, with their gossamer-thin fibers, that gave them their special properties and accounted for their prodigious success.

The origins of their success can be dated to a dinner party in 1970. Japanese designer Issey Miyake, the story goes, showed up in a jacket made from a faux suede that had recently debuted at a Paris fashion show. Seeing it up close was another designer, Halston, born Roy Halston Frowick, in 1932, in Des Moines, creator of the pillbox hat Jackie Kennedy wore at her husband's inauguration in 1961. "I flipped," he told a reporter for the *New York Times*. The material felt soft, looked luxurious, left him wondering what he could do with it.

What he did with it was design a dress with buttons down the front like a man's shirt, known as a shirtwaist dress. "It started as a blouse," he'd tell the fashion tale, "developed into a long evening gown and ended up as a casual shirtwaist. I made over a hundred sketches before I got what I wanted," namely his Design #704. "That single dress," a coffee table biography of Halston observed, "became a uniform for the Upper East Side set." He did it in 36 colors. "That's more than Baskin-Robbins does for ice cream," he said in 1976. Then came the safari suits for men and women, the skirts, culottes, battle jackets, and trench coats, often in cool camels and grays. And perhaps the greatest part of their allure lay in the material from which they were made.

Its Japanese developer, Toray, at first gave it the code name Toray-223. Later, it went by the name Ecsaine in Japan. In Europe, after Toray collaborated with an Italian firm to make it in a factory in Terni, Italy, it became Alcantara. An executive of its American distributor, Springs Industries, gave it its name in the United States, Ultrasuede, which is what I'll call it here. Ultrasuede felt soft and drapy, much like real suede. It didn't wrinkle. It could be thrown in the washing machine. And it was strong. But there was nothing grimly functional about it; like leather, it looked and felt luxurious. Even grainy photos of Ultrasuede-garbed models strutting down catwalks during the 1970s convey its easy flowing softness. In October 1974 a smiling Princess Grace posed on the steps of the palace in Monaco, regal and radiant, her hand brushing the pocket of a floor-length dress made of Ultrasuede. By 1976 the American fashion industry was annually using 1.5 million yards of it. In the years since, it's had its ups and downs. Not everybody likes it. Some object to its distinctive rustle; it makes "a kind of whopping noise," complains one fashionista. But it has remained a persistent market presence, available in a host of weights, special treatments, and colors, its name in America now almost a Kleenex-like generic for faux suedes generally.

"Did Halston make Ultrasuede or did Ultrasuede make Halston?" asked Elaine Gross and Fred Rottman in their biography of the designer. "In the 1970s, the two were synonymous." Other designers had initially dismissed it as just another inferior synthetic but in the end also embraced it. According to Gross and Rottman, Halston's success rested on his insistence that Ultrasuede be used like suede, not fabric. You didn't need linings or interfacing; like suede it was substantial enough on its own. You'd forswear regular seams because in Ultrasuede they'd be too bulky, couldn't be pressed to lie flat; you'd sew it up with overlapped edges and topstitching instead. Polyester and polyurethane were what went into it, but "unless you happen to be a chemist," a *New York Times* article said in 1976, "do you care? What interests women about Ultrasuede . . . is how close it comes to looking and feeling like the real thing."

In 1984 the iconoclastic Texan political essayist Molly Ivins indulged in a wild reverie in the *Dallas Times Herald* about trying—really *trying*, she tried to make it sound—to understand Republican women delegates to the party's Dallas convention. Oh, they were united, of course, by "the essential Republican ethos"—comfort with what they had and determination to keep it. But the species under her microscope, it seemed to her, came in two varieties. First, those with more money, not quite so conservative, and embarrassed by the party's evangelical wing; these were the Ultrasuedes. They, in turn, fretted about encroachment of the second type, a lower sort you might want to keep out of your country club, the polyesters.

As usual, Ivins was poking fun and having fun. But if you wanted to get picky about it, and maybe a wee bit over-technical, you'd have to say she got it wrong. High-toned Ultrasuedes versus low-life polyesters? Ultrasuede *is* polyester. Not exactly, and not all of it; the label says 60 percent polyester, 40 percent polyurethane. But the polyester fibers going into it are, chemically speaking, identical to those found in those tawdry, sadly maligned leisure suits that gave the 1970s, and polyester itself, a bad name.

Even when chemically identical, however, all polyester is not alike. A guitar string is not a bridge cable, though both might be made from similar steel. The old polyester fiber of unhappy memory was coarse; that in Ultrasuede was exceedingly fine. And that difference makes all the difference.

It is 1986, and Joe Rivers, years past his early Corfam work with John Piccard, is being interviewed by David Hounshell, a historian of technology gathering material for a book about Du Pont R&D. One influence on the properties of synthetic fibers, Rivers explains, is viscosity, the globbiness of the molten plastic fiber. Too low, and it "goes through the spinneret like water." No good.

What else affects its properties? asks Hounshell.

"The denier," says Rivers. "If you can get the filament deniers down," he explains, "then you get a softer, more pliant yarn."

Denier, pronounced DEH-nyer, is the traditional gauge of fiber or yarn thickness, the word going back to a silver coin of that name used during the middle ages. Weigh, in grams, a standard length of fiber—9,000 yards, about five miles—and that's its denier. A coarse, scratchy wool fiber might be 15 denier, cotton or polyester 3 to 6, silk around 1. That's pretty fine; by definition, 9,000 yards of 1 denier silk weighs just 1 gram; you could mail 100 miles of it with a single first class stamp. As Rivers was explaining to Hounshell, the lower the denier, the softer—in fact, *much* softer. Because as the formulas in any structural engineering text imply, one filament a tenth the diameter of another is only one ten-thousandth as stiff. So, make a synthetic leather with wraithlike, low-denier fibers and maybe you could get something as drapy, soft, and seductive as some silken garment leather.

But how, as Rivers put it, do you *get the filament deniers down*?

One way to make synthetic fiber is to start with molten plastic and eject it through a tiny orifice, like water from a water pistol; almost instantly it cools and solidifies, whereupon you can wind it up onto a bobbin. Another way is to start with the fiber-forming material in solution and extrude it, through a little hole, directly into a coagulating bath; a jet of solid plastic forms as it passes through. A third is to have hot air evaporate the solvent as it's shot through the hole, leaving the solid filament. Fibers made by the first process are called melt spun, by the second wet spun, by the third dry spun. In each case, you start with a liquid, shoot it through a hole, and get a continuous filament.

The contrivance with tiny holes that forms the fiber is known, from the early days of rayon, as a spinneret, named for the organ in the silkworm that "spins" silk; it has nothing to do with spinning filament into thread or thread into yarn. Nylon fishing line? It comes from spinnerets; the very first used to make nylon in the lab, Du Pont likes to say, was a drugstore hypodermic needle. When they made Corfam, the polyester fiber later cut into short lengths of "staple," first came through a spinneret. Visit Du Pont's Experimental Station today and chances are you'll see some experimental rig with a fine, spidery line of glistening polymer filament emerging from it. Spewing out the nylon, the polyester, and acrylic of the world's textiles industry are banks of these precious little things, expensive showerheads, lovingly wrought from platinum, or fine stainless steel, each with lots of holes. The smaller the holes, the finer the fiber.

Finer, that is, up to a point. Make the holes too small, draw polymer through them too aggressively or too fast—and spinning speeds can run upwards of 3,000 meters per minute, 100 miles an hour—and ultimately, no matter how tough the material, it will break. So there's some lower limit to how fine these tendrils of plastic can be. For polyester it turned out to about 0.3 denier, somewhat finer than silk. But if you wanted yet finer fibers—like, say, the ineffably fine collagen fibers of leather? Well, you couldn't get them.

And then Miyoshi Okamoto did get them.

Miyoshi Okamoto was born in Kouka, the old ninja capital of Japan, in 1936. In 1960 he graduated from Nagoya Institute of Technology with a degree in industrial chemistry and went straight to work for Toray, at a lab in the company's old Ehime plant, on the island of Shikoku in western

Japan. Rayon, on which Toray had grown up, was on the decline. Acrylic and polyester were the materials of the future. Okamoto was soon assigned to the Mishima polyester plant, where he stayed in each department for a few weeks to give him raw-materials-to-final-fiber experience. Along the way he was assigned to come up with a new, bulkier polyester.

Going up in Japan were more Western-style buildings whose cold hard concrete floors traditional tatami mats never seemed able to soften enough; needed were more materials like tufted carpet, the softer and bulkier the better. Okamoto conceived a way to produce "ultra-bulky" polyester by spinning it in two slightly different formulations, through special spinnerets that ejected them side by side. Select the two on the basis of unequal shrinkage, and the conjoined fiber could be made to crimp up all you like. Soon the company was making 6,000 tons of it a year, making for softer, flouncier carpets and futons. Okamoto was a star, after just three years with the company, recipient of a big bonus and buckets of praise, and all of 27 years old. What next?

His bulked-up polyester was a composite fiber, its properties depending on more than one component. But Du Pont was already the acknowledged leader in this field; everyone else, he wrote in a Japanese-language research memoir published in 2005, was "jumping into the fray and playing catch-up." A dead end if ever there was one; he wanted something new, something "Du Pont would never even *dream* of." It was 1965, right after the Tokyo Olympics. Transferred to the company's central research laboratory in Shiga, a big early success under his belt, he felt he could let his imagination roam. He combed the patent literature. He explored different sorts of spinnerets, the heart of fiber spinning. And, to hear him tell it, he looked to nature.

Growing up near Kouga, about 40 miles north of Tokyo, Okamoto had felt an affinity for its mountains and thick forests, heard stories of ninja warriors who expertly melted into them, hidden by smokescreens of burning grass. He saw his father develop home remedies from local crops and sell them to *baiyaku*, or itinerant medicine peddlers. Stricken by illness as a child, Okamoto had recovered thanks to penicillin, made from the green mold discovered by Alexander Fleming, at whose grave in London he later offered thanks. "Nature," he'd grown up to believe with unusual fervor, "is the best teacher." If he was to develop new synthetic fibers, what could he learn from natural ones?

Silk was among the most desired of the natural fibers—and the finest. Wool? That of the vicuna—today it can cost $225 a pound—was most coveted, its fibers more delicate than those of sheep's wool or cashmere. Paper? The best of it, used to manufacture money, came from the fine fibers of kouzo and mitsumata bark. Leather? After tanning you were left with branched and entangled collagen fibrils of unimaginable fineness. What people valued most in natural materials, it struck him, was this lightness and delicacy.

Okamoto noticed, too, that putting natural materials to human use often meant extracting something from them first. Silk consisted of fibroin fibers bound with sericin, which you had to remove. Making leather meant stripping from the hide almost everything but those microscopic cables of collagen. *Remove extraneous substances. Wind up with something desirable and fine.* Maybe that was a principle worth pursuing.

In 1965, he took what his superiors at Toray might otherwise have written off as an empty idea, hopelessly abstract, and transmuted it into something real. Working with a single assistant, he gathered lengths of nylon filament and polyester filament, mixed them, and packed them into a fat glass test tube, an inch in diameter, 8 inches long; imagine it stuffed with a bundle of two different kinds of uncooked spaghetti, semolina, say, and whole wheat. One end of the test tube had been formed into a funnel culminating in a sharp tip; Okamoto broke it off, forming a tiny nozzle. At the other end, just a plain cylindrical opening, he fitted a wooden piston. In the middle, between piston and nozzle, went the polymeric spaghetti mix.

Then he immersed the apparatus in a hot silicone bath, kept at 300 degrees Celsius. That's just above the melting points of polyester and nylon—enough to first soften, then melt them both. Next, he gently pushed the piston, squeezing the now molten fibers, extruding them through the nozzle. As he did so, the inch-thick fiber bundle emerged as a filament just a few thousandths of an inch across—but with the original lines of spaghetti, each now microscopically fine, emerging intact. Instantly solidified by a flow of cool air, the filament was then wound up onto a bobbin by an electric motor. The result reminded Okamoto of *Kintarou-ame,* a popular candy with just such fine striations running down its length.

Then, Okamoto wrote later, "I stretched out the fiber, under heat, to three or four times its original length and looked at my creation under a microscope." As it lay open to him on the glass slide, the fiber consisted of about 70 percent nylon strands, 30 percent polyester. Of course, it almost didn't matter just what they were, so long as they were different. For now came the step toward which all this was leading, one designed to act on the two polymers unequally: *Formic acid dissolves nylon but not polyester.* "On a glass slide, with a drop of formic acid," Okamoto wrote later, "a single strand of fiber turned into what must have been tens of thousands of polyester microfibers"—the nylon was gone now—"shining like pearls." *Funya-funya* and *tara-tara* are the Japanese words Okamoto would use to describe these glistening microfibers; the first means soft and pliable, the second supple, with an easy drape. With this step it was as if the thick ropelike pigtail of some Teutonic Brunhilde had sprung loose into the flowing silken hairs of a young girl.

This wasn't how they'd ultimately make Ultrasuede, but the principle—dissolve out one component of a fiber, leaving many fine microfibers behind—was established. The demonstration convinced Toray higher-ups to free up funds for the expensive spinnerets and other technology that became the focus of the next stage of work. "Not all of this proceeded smoothly," writes Okamoto. Their prototype spinning apparatus had to run continuously, day and night; he and his few assistants faced power outages, lightning strikes, and equipment failures. "Before long, I was making nightly pilgrimages" to a local shrine after work, "to pray for protection from bad luck."

Until now Okamoto's research was what he has termed in English "discovery-driven"—roughly, basic research; no uses for it had yet been established. He'd shown he could make the extremely fine fibers that, to his mind, held promise. Needed now were concrete applications, something for the company to make money on. Perhaps, in silk-loving Japan, a synthetic silk better than rayon or nylon? Or a synthetic leather or suede? Texturing, weaving, dyeing—any such later treatments would depend on its final use. In a fever, Okamoto cast about within the company for possible applications. Synthetic paper? Fur? Wool? He was forever off to other divisions and departments, talking, listening; seeing his desk so often empty, he suspects, his colleagues must have thought him lazy or irresponsible. "I visited labs one after another, but I never stayed long enough to warm the laboratory benches," is how he put it later. Today, nearing 70, showered with honors, he comes across in photos as the imposing lab director, the distinguished senior figure. But a shot of him from much earlier, while still in his 30s, standing straight and tall in the frame of the photo, makes it easier to imagine him all youthful urgency, navigating the bureaucratic shoals of Toray.

And one among his round of visits did pay off.

In 1966, Hitelac was foundering. Hitelac was a Toray synthetic leather that seemed to Okamoto a copy of Corfam. Aimed at the footwear market, it was hard and shiny, with an unappealing hand, and made for shoes that, as Okamoto writes in an email, "bit a person's feet." *Hard*—that's what was the matter with it. And the fibers from his glass cylinder experiment, whatever else they were, were soft. "This rang a bell for me," he wrote. A friend who'd joined Toray the same time he had, Makoto Kounosu, was working with artificial leather and "struggling with the stuff being too rigid." Pretty soon the two of them made up a sample of artificial suede based on Okamoto's early microfibrous material. It was about the size of a notebook and brown. Or rather its pigmented binder was brown; they hadn't bothered to dye its white microfibers. By later standards it was a sad affair, but it was supple, and supple was what suede needed to be. Today, their original sample, a little ragged with use and wear, is on display at a company research center in Mishima, along with the scissors Okamoto used back then to slice up samples.

For a while they pursued both smooth leather and suede. Okamoto and his colleagues didn't feel quite at home with leather, which wasn't big among the Japanese; the appeal of a napped suede surface, ordinarily so attractive to Westerners, was to some of them a mystery. Meanwhile, as Okamoto writes, "Applications for artificial shoe leather were a difficult, specialized world for us." At one point they asked a visiting foreign consultant for his opinion. "He made an exaggerated gesture of throwing away a shoe leather prototype, but then rubbed a suede-like" one against his cheek, praising its incomparable softness. Too bad his bosses hadn't witnessed the little show, thought Okamoto; hearing about it from him they'd write it off as just another researcher head over heels in love with his own project. But that one gesture, etched in memory, helped convince him that in suede lay the path ahead.

He and his colleagues began taking out patents on the technology. In a key early one, filed in Japan in 1966, Okamoto and co-workers sought protection for what would be called, in its 1970 U.S. counterpart, "Synthetic Filaments and the Like." There they described a filament formed from two polymer elements, one of which "is distributed as islands in the sea of the other polymer element when seen in cross section." When, a little later, they patented a mechanically intricate spinneret to make it, the patent was entitled "Apparatus for Spinning Synthetic 'Islands-in-a-Sea' Type Composite Filaments."

Imagine a map of a South Pacific archipelago, its diminutive island dots set against the ocean vastness. Well, if you sliced off the mixed spaghettis of Okamoto's original experiment and viewed them head on, a Pacific map like that is about what you'd see—dark "island" fibers coming up out of the paper surrounded by lighter ones, collectively the "sea." Design your spinneret right and you could have, as subsequent refinements of the technology actually did, 145 island fibers issuing from a single islands-in-the-sea complex, each wrapped in protective sea.

Indeed, this was a key virtue of the technology: All the way through manufacturing, the fiber could be as thick, and robust, and easy to work as you'd want—in one case, reflecting later practice, 7.5 denier, like coarse cotton fibers. So you'd have a tough, thick-fibered web when you needed it to be tough and thick fibered, during manufacturing. It was only at the end, once you'd dissolved away the rigidifying sea, that you got the lush, silken softness of the final material.

The beauty of it all came down to two simple sketches, worthy of a child's drawing, lying side by side in one of the Toray patents. One shows a few meaty islands-in-the-sea fibers caught up in one another, before you've done anything with them. The other shows the same fibers, once you've dissolved out the gluelike sea that holds them rigid, each now a bundle of inexpressibly fine microfibers. A tangle of them would have something of the suppleness Okamoto had glimpsed in his first experiment, the denier of each no longer 6 or 7 but a tenth that, or less.

Or *much* less. Toray has since claimed polyester fiber of 0.00009 denier; fiber like that weighing only as much as a restaurant packet of sugar would reach to the Moon.

For three years Okamoto had been working on the technology, half a dozen assistants at his side, when in the spring of 1968 the project came up for review. He brought with him his sample of brown faux suede, along with an account of progress over the past three years aimed at scaling it up for production. Result? His work was "excoriated." That's the word his translator uses for Okamoto's characterization of his treatment. Company vice-president Yoshitsugu Fujiyoshi concluded it would cost too much to make the stuff. Potential applications were too few. Research support was withdrawn . . . but not entirely. "I still had some behind-the-scenes supporters," Okamoto writes. If they were to simply stop, his sympathetic boss told him, it might seem there'd been nothing much to the idea in the first place. Forge ahead, and the company would ultimately come around. So the work continued, if more slowly.

Around this time Toray got its corporate hands on one of the new scanning electron microscopes, which allowed you to get high-resolution images with features down to a few ten-millionths of a meter. You've seen them, the scary, blown-up photos of bugs that make them look like monsters from Mars; those are typically SEM images. Okamoto wasn't interested in bugs, of course, but in materials like wool, cashmere, paper, fur, feather, silk—and, especially, suede, made from deer hide that, in the SEM images, vividly revealed its fibrous fuzz. This new microscopic power let them better see what they were doing. And let them better *show* what they were doing to company officials. "Success in developing artificial leather," Okamoto and a coauthor would write in a 1998 review of *shingosen*, the specialty fibers of exquisite touch and feel that Toray and other Japanese companies pioneered in the years after Ultrasuede, "is much due to the microscopical observation of real leather and the efforts to mimic" its structure. Ultrasuede, Toray would say, was "What Nature Had in Mind."

Early in 1969 Okamoto was named head of a new group in the artificial leather business that Toray hoped to resurrect; it was called the Second Business Unit, he writes, "at a time, when we had no first business unit!" For nine months his and two other groups of researchers, culled from Toray labs and departments, competed—or rather, as he wrote in an email in English, "were in a kind of severe competition state." His group of five, pursuing microfibers, was by far the smallest and, in the wake of the research review, least favored—a David slugging it out with two Goliaths.

In fact, microfibers were still mired in problems. One was dyeing them; they were so fine, their surface area so large, that typical dye concentrations left anything made from them look washed out. One day Okamoto forgot to turn off a Bunsen burner on which a beaker of dye simmered. The solution turned thicker, the color deeper, and he was led to suggest dyeing at higher concentrations. That, he learned, would work better with polyester as the "island" than with nylon, which they'd also been considering; so polyester it was. Another problem they faced in those early months of 1969 was the needles used, as Corfam had, to make nonwoven felt; all you could get just then, Okamoto remembers, was "low-grade rubbish covered in needle marks"; that would never do for the fine garment suedes they envisioned. He searched for finer needles and the technology to use them. Later, when a long-dubious equipment manufacturer with whom they were working "watched [nonwoven] felt of unprecedented beauty coming out of the machine, he fell utterly silent."

Finally, in September 1969 the phone rang at the Second Business Unit. Toray executives visiting America, armed with samples of the prototype faux suede and SEM images comparing it to real suede, phoned with the first good news. More phone and telex messages followed. "We were to develop production technology within six months!" His boss took the first call, but Okamoto was right there for the moment of triumph. Company resources poured in.

The following year, at a haute couture fashion show in Paris, several designers showed clothes made from the new faux suede. This "touched off a sensation," wrote Okamoto, applause for the garments garbing the Paris models at first refusing to subside. Of course, "back in Japan, the event merited only a three-line description in a newspaper article. That's how unfamiliar Japan was with leather at the time." The Paris hooplah didn't translate instantly into business, but then the orders did start coming in, no doubt after Halston, whom Okamoto later met, created his signature shirtdress. "The Rolls-Royce of fibers," someone called Ultrasuede. "The most revolutionary clothing material since Adam and Eve's fig leaf."

It was also called, Okamoto said later, "'the biggest invention since nylon.'"

❧

A brief video devoted to the manufacture of Ultrasuede, presented in his New York office by Yoshifumi Hamada, president and CEO of Toray's American division, offers a stream of images straight from some indeterminate border country between photographic realism and Hollywood animation. We see fibers cut into staple, nonwoven fluff grow denser, "sea" material dissolved out. But it's so fast, so full of color and aesthetically blurred images, that the effect is more smoke and mirrors than nuts and bolts. By one reckoning, 250 patents protect Ultrasuede, 100 of them in the United States—patents on web forming, needlepunching, dyeing, and spinnerets. And while it's been nearly 40 years since its introduction, most production details, Traci May-Plumlee and Thomas F. Gilmore wrote in a nonwovens industry overview, remain trade secrets. "Because the technology was so revolutionary at the time," they suggested, "Toray was able to acquire a patent which was very broad, avoids specificity, and consequently sheds very little light on the specifications for creating Ultrasuede."

Ultrasuede would be pictured as a textbook case of personal creativity and corporate innovation. Okamoto, it was said, benefited from *angura,* the underground research policy that let Japanese scientists spend time on their own projects, off the radar of higher management. Had Toray's executives been more narrowly dollars-and-cents oriented, an industry analyst was quoted as saying, "they probably would have given up. And if they strove for harmony like other Japanese companies, they probably wouldn't have succeeded." In 1989 Toray named Okamoto a director and honored him with a lab named for him, specifically intended to encourage research independence. "We don't even ask what they're doing," Toray's R&D director said. "To make researchers more creative, you have to let them screw up." Later, after Ultrasuede started being produced in Italy under its European name, Alcantara, Okamoto would be named recipient of the Leonardo Prize, for contributing to the global luster of MADE IN ITALY. A story told later pictured Ultrasuede as product of "a spinning technique invented by Miyoshi Okamoto while meditating on a whirlpool," a story he doesn't deny.

Notable as it was, however, Ultrasuede was only the most successful of several microfiber-based synthetic leathers introduced in the 1970s. It was not alone to be based on two-component fibers. It was not by every yardstick even the first. Okamoto himself credits Du Pont innovations in bicomponent fibers. And Amanda Lindsay, a young British doctoral student, reported in her 1999 thesis, "The Evolution of Microfibre Through Technology and Market Pressure," that one early vision of two-component microfibers went back at least to 1939. A British patent filed that year by IG Farben, the German chemical cartel later implicated in war crimes, suggested how "artificial fibres of any desired fineness" might be made from two different liquids that didn't mix and whose properties differed. If they came out of the spinneret like two coaxial fibers, you could dissolve away the inner core and be left with a hollow fiber. Or else you could dissolve away the outer cylindrical layer and wind up with a much finer fiber—though apparently nothing less than 1.5 denier was achieved. Moreover, as Toray later did with Ultrasuede, you could wait until you'd made the fabric before dissolving away the second component. All this was envisioned by the IG patent, details of which, Lindsay found, appeared in a British textile handbook in 1946.

These were not the only ways in which Ultrasuede, like many genuine innovations, was not in every way original. Indeed, Toray—as did most of its Japanese competitors—in some ways only aped what Du Pont had done with Corfam. Like Corfam, Ultrasuede used a nonwoven substrate. Like Corfam, it was needlepunched into a tight web. Like Corfam, it used as its binder a polyurethane; this proven class of polymers could be formulated to give a hint of elasticity, to hold the straying fingertip the way leather did, and, unlike vinyl, to neither turn sticky when hot nor crack when cold.

By 1983, 11 million meters of microfiber faux leather were being produced in Japan each year, about a third of it Ultrasuede, the rest in at least 10 other products. These days Ultrasuede has yet more competitors, some in Korea and elsewhere in Asia, as well as Italy; at one point Toray Ultrasuede (America) announced plans to help consumers "distinguish Ultrasuede products from its imitative competitors." The consumer in the mall might be oblivious to the differences, but differ they do. Ultrasuede is 60 percent polyester; Amaretta, made by Kuraray, is 60 percent nylon. In Ultrasuede, as we've seen, spinnerets fashion the islands-in-the-sea structure directly. In some competitive materials, fiber elements come out of the spinneret weakly linked and then are shaken free; it's a little like a novelty chocolate I like, shaped like an intact orange that, smartly tapped, breaks cleanly into orange sections. Some from among these many offerings are imitation suedes. Others, like Clarino, bear leather's smooth iconic grain. Still others, like Lorica, can be had in either. A conscientious student will appreciate the charts, tables, and technical specs needed to properly keep track of them. But all are based on collagen-mimicking microfibers.

"For a long time," began an American patent for an early faux leather, first filed in Japan by Kuraray in 1963, "the manufacture of synthetic articles having the same composition as that of natural leather has been a dream for the skilled in certain arts"; with this invention, it so much as said, that dream had at last been realized, the result a material that "bears a striking resemblance to natural leather in structure, external appearance, texture and touch." The inventors were Osamu Fukushima, Hiroshi Hayanami, and Kazuo Nagoshi, assignors to Kurashiki Rayon Company Limited, Kuraray's predecessor. In January 1969 they were awarded U.S. patent 3,424,604. It's just one of dozens granted Japanese synthetic fiber makers beginning in the 1960s that helped establish their dominance in synthetic leathers. But in its earnest review of past failures, its evidence of over-hasty translation from the Japanese, and in sometimes sounding more like manifesto than patent, it lends our story a charming, even idiosyncratic, note.

One early process took leather scraps and crushed them, sometimes adding some appropriate fragrance as "camouflage." This, said the Kuraray inventors, was "regenerated leather."

Next came "imitation leather," such as those coated "in the early time" with nitrocellulose, vinyl, or rubber.

Then came "'synthetic leather,' which exceeded the stage of 'imitation leather,'" and used nonwoven fabrics and microporous binders; here was Corfam and its poromeric ilk. These, too, however, being "so inferior to the natural leather," had spurred demand for even better leathery lookalikes. In Japan, shoe manufacturers demanded, but did not yet possess, "a synthetic leather with an external appearance and touch similar to those of natural leather."

From here it was on to a clear-eyed recitation of leather's virtues. The first was a "unique beauty attributable to its grain surface. Commonly the expression 'leather-like' has been used, and leather-made ornaments match well both Japanese clothes and European clothes." Leather was pliable, moisture absorbing, easy cutting and lent itself to dyeing—nine specific advantages, all told—though its real strength lay not in any one of them but in "the excellence of the combination of the individual properties." And so on for leather's shortcomings.

This editorial preamble went on for 28 paragraphs—here was "doing your homework" with a vengeance—with no mention of the invention itself. But finally, with the herculean obstacles faced by the inventors duly recounted:

> Accordingly, it is supposed to be a task for one who intends to make synthetic leathers to eliminate such defects as found in the natural leather, to improve such advantages as in the natural leather without any depression in the other properties and, thereby, to improve the combined properties.

It was the "object of the present invention" to do that, the claimed result being a "pliant, supple, air-and-moisture-permeable sheet material" the equal of leather.

The patent described a "mix-spun" fiber consisting of two polymers mixed together and expelled through an ordinary spinneret, into a comfortably thick fiber, maybe 3 denier, tendrils of one polymer worked into the other. Once the fiber was formed into a nonwoven web, you'd dissolve out one component, being left only with the other. Depending on which you dissolved, you'd get fibers either riddled with pores or bundles of fine filaments. Either way, with so much of the original material washed away, the fibers would be of lower denier, in some formulations around 0.5. The result? A pliable, fine-fibered mat. No need for the special spinnerets Okamoto and Co. would use for Ultrasuede. But thanks to its own islands-in-the-sea qualities, it left you with similarly fine fibers, making for a similarly soft and supple faux leather.

Thirty-five years after the award of that patent, Kuraray was all over the synthetic leather map, its microfibers used for gloves, basketballs, and shoes and as raw material for Lorica. Since the 1960s, Kuraray has made Clarino, first in a form Du Pont could write off as inferior, later much improved with microfibers. Today, says Atsushi Kumano, marketing manager at the company's New York office high over East 52nd Street, they cover softballs and line jewelry boxes with it and have tried to convince a Texas bootmaker to use it, perhaps even for saddles. But while Clarino competes with leather, it approaches Du Pont's goal for Corfam, that it simply be accepted for itself, not as faux anything. Says Sanchiko Barbasso, Kumano's assistant, of Kuraray's longest-serving synthetic leather, 40 years in the market: "It's just Clarino."

Like Toray, Kuraray has a seductive website where it shows off Clarino and the company's next-generation offerings—such as, most recently, Parcassio; by one dollop of hype, this is "a virtual 'man-made cloning' of natural leather," built on two microfiber layers fused together, coupled to Korean and Japanese tanning technology; one company using it for shoes promoted its "broken-in comfort, right out of the box." Then there's Amaretta, with the same nylon-polyurethane formulation of its Clarino antecedents but thinner, drapier, competitive with Ultrasuede and lambskin and available as a smooth-grained synthetic garment leather as well as suede.

A few years back, hoping to make Amaretta better known in the American fashion marketplace, Kuraray approached Mira Zivcovich for her help.

Mira Zivcovich's studio, up three fluorescent-lit flights of bare metal staircase from West 14th Street in lower Manhattan, is heralded by a red door flanked by framed samples of her advertising agency's work. Once inside, the place opens up into a large open area set off by bookshelves and screens, warm and homey. This is Mira Zivcovich's creative lair. In 1966, as a teenager, she came to the United States from Belgrade—her father was a filmmaker and photographer in the former Yugoslavia—and by the 1980s she'd set up her own ad agency.

It was around a conference table at Kuraray's office in midtown Manhattan that she and an assistant met with company officials armed, she remembers, with "hundreds of pages of scientific papers" about Amaretta. The idea was to capture the attention of top-end designers, people who didn't much think or care about material going for less than 50 bucks a yard. She didn't think tech specs were the way to do it. How then? About all Kuraray could say for sure was that they wanted to set off Amaretta from its "aesthetic competition"—that is, Ultrasuede; anything she came up with, she was told, had to point up Amaretta's superiority to Ultrasuede, by name.

Zivcovich gathered testimonials from New York designers on behalf of Amaretta. But, especially at the beginning, she and three colleagues mainly brainstormed, tapping into the computer taglines, Q&As, slogans, and scenarios, unconstrained at first by spelling, typos, or the dreary exactitude of facts:

> Designers want Amaretta. Designers need Amaretta. Designers love Amaretta. Even if they don't know it yet.
>
> *
>
> It looks like suede, but feels like velvet.
>
> *
>
> [Its] three-dimensional structure is idential to that of natural leather, but superior in quality.
>
> *
>
> Wearing Amaretta is almost like having a super sexy second skin.
>
> *
>
> Lamb Skin? Baby Pig Skin? Not by the hair on my chinny-chin-chin!
>
> *
>
> Love at first touch.

And of course:

> The old days of suede and leather are over.

In one over-the-top, night-on-the-town scenario they dreamed up, the Amaretta jacket you've slipped across the shoulders of your date falls into an oily curbside puddle. A cab driver plows over it. A pill-popping nightclubber—"his girl friend left him for a Euro Rock star"—vomits all over it. No problem, though, not for Amaretta. Just throw it in the washing machine. "This miraculous mother of all inventions is also machine washable."

All this, of course, was just playing, a wide-open search for ideas. The final eight-page brochure, "Amaretta: Creating Miracles from Nature," mostly in blackberry purple and fire engine red, showed a three-dimensional artist's rendering of Amaretta's "dense matrix of super-thin nylon fibers surrounded by billions of tiny pores." It boasted that Amaretta breathed yet repelled water, that its microfibers were more micro than those of Ultrasuede. It showed beautiful male and female models, their nakedness covered not in finished skirts and blouses but raw bolts of Amaretta, suede and smooth grain.

One panel showed two such golden-tressed treasures, she frizzy-haired, bronzed, and braless, he a pretty boy with long locks swooping down fetchingly across his eyes, a rocky precipice barely visible behind them. But, oh, there is a third figure in the composition. The young man has one arm around the girl, the other around a sheep.

AMARETTA

NATURE'S BEST FRIEND

"The strongest synthetic leather looks and feels like the real thing," says the caption, "without sacrificing man's best friends."

It was quite a trick pulling off *that* shot, Zivcovich recalls. Picking her way through the natural versus synthetic minefield, she'd not wanted to offend designers who cherished leather but *did* want to make the point that while it "looks and feels like the real thing," as the ad copy said, Amaretta did no damage to nature. . . .

A lamb. She'd use a lamb to make the point.

But they couldn't *get* a lamb, so they had to manage with a sheep—a little bigger, not quite so cute.

Only thing was, sheep were barred from Central Park, where they hoped to shoot; dogs were OK, but not sheep. Undaunted, Zivcovich had an upstate farmer smuggle one in by van. "The sheep cost more than the models," she says.

But then one of the models needed calming down; he was afraid of the sheep, and getting him to look at ease in front of the camera, for even the fraction of a second needed for the exposure, took some doing.

In her brochure, Zivcovich went beyond simply asserting that, by some technical or aesthetic standard, Amaretta was as good or better than leather. She pointed to a sometimes forgotten truth of leather—that it was the skin of a dead animal, and Amaretta was not.

NO DAMAGE TO NATURE, OR ANY LIVING CREATURE.

10

Crocodile Dreams

i. "My Skin's Not For You"

We are 30 blocks south of Central Park, on East 26th Street, border country between higher-profile Manhattan neighborhoods, a little nondescript. On one side a shop offers $10 pedicures. On the other is a parking lot, its chain link fence topped with razor wire. In between, on the site of a defunct butcher shop, is Moo Shoes. A cat lounges on its round bed. Customers tiptoe around it. Out front a sign in bold black-and-white hide markings tells you Moo Shoes offers:

ALTERNATIVES TO LEATHER

The Kubertsky sisters, Erica and Sara, started Moo Shoes here in 2001. (It's since moved to the Lower East Side.) They were vegan; they forswore animal products in their food and didn't want them on their feet, either. "We just wanted to know," they say, "that those hip, classy, good-looking, well-fitting shoes were leather-free." Now, belts and handbags, as well as men's and women's shoes, crowd the store's galvanized steel shelves. A pair of black stilettos from Portugal for $65. A strappy Rinaldo with coppery faux surface. Birkenstocks. A vinyl belt. They're fake and proud of it.

REAL FAKE
GENUINE LEATHER-FREE GARMENT
MADE IN ENGLAND

Here, within this little shop, the world is turned upside down: Here it's "leather-free" that's genuine and coveted, not leather—bags, books, wallets, and shoes that are "cruelty-free."

When Mira Zivcovich staged her photo shoot on behalf of "Amaretta: Nature's Best Friend," the sheep smuggled into Central Park that day was meant to stand for innocent animals its use presumably spared. These days even Naugahyde gets into the act: the company website says its zany fictional Naugas yield fabrics that look like leather. But "since Naugas shed their hydes without harm to themselves," Naugahyde is "the Cruelty-Free Fabric." (What the company doesn't want you to know, one web posting smirks, twisting the joke one more tasteless turn, is that "in the spring they go down to the river and club Naugas on the head, skin them, and turn them into fake leather.")

The energy to overturn the old sacrosanct hierarchy of leather at the top and plastic below comes from PETA, People for the Ethical Treatment of Animals. In a 2003 *New Yorker* article about founder Ingrid Newkirk, Michael Specter called PETA "by far the most successful radical organization in America." Animals bred and slaughtered solely for the value of their fur? Meat eating generally, the sin only made worse when animals are kept indoors their whole lives? Circuses in which bears and elephants are cruelly trained and exhibited for human pleasure? Doesn't matter. PETA's singleminded message brooks no compromise: Animals are sentient beings. They need to be treated ethically. So long as humans exploit animals, PETA objects. It campaigns against animal experimentation. It sells cruelty-free mousetraps. It lambastes insensitive advertisers. It protests the use of leather.

"Leather is just hairless fur," its website proclaims, setting out the arguments against it. A mere by-product of the meat industry, as the leather industry contends, the animal is, after all, already dead? No, says PETA, "the animals are dead *because* there is a demand for their flesh and skin." The hide—to PETA it's always *skin*, like yours or mine—contributes to profit. "Every time you choose to buy a leather jacket or leather shoes, you sentence an animal to a lifetime of suffering." Alternatives? There's no want of them. "Make your closet a cruelty-free zone," PETA advises. Its 28-page "Shopping Guide to Nonleather Products" lists vegan companies, including Moo Shoes. Synthetics, it insists, breathe and "give" as well as leather or better. Pollution? Tanneries are worse than most "chemical" plants.

As part of a PETA web campaign called "Shed Your Skin," Pamela Anderson—pitched as "the sultry *Baywatch* star," all eyeshadow, blond tresses, and pouty lips—breathily intones:

> You might not think about it every time you get dressed, but every time you put on a pair of shoes, a belt, or a pair of gloves, chances are you're wearing leather. And of all the things animals are abused for, leather is the most common.

Anderson tells a sorry story of cattle trucked to awful deaths in India, their hides becoming leather goods bound for Nordstrom, Gap, and other retail outlets. In a street interview in New York City, a young man in a watch cap, still wearing leather, seems to come around on the spot: "There comes a step in evolution where we fabricate clothes, synthetic fibers. We can fabricate food. We have the means. It's just time to do it." In our age, says PETA, we don't need leather. It's a vestige of the ignorant past, the pain it inflicts unnecessary. That it's "natural" counts for nothing.

PETA mightn't be so effective were it just another earnest schoolmarm of an activist organization. But while it's nothing if not serious, it relies on catchy phrases to deliver its message, on madcap ads, memorably bad puns, and testimonials from pop culture icons like Paul McCartney. The use of leather, as PETA tells it, is cruel and unethical, yes, but also very, very uncool.

In an award-winning PETA public service announcement, old black-and-white footage shows cows, steers, and calves chomping contentedly away, their mouths synched to loony, contagiously twangy country music:

> Our sweet backsides
> Would like to stay together
> We don't want to be your leather.
> Won't you show us some love,
> Don't you make us kid gloves,
> Or a jacket or a shoe,
> Cuz my skin's not for you.
> Do I make myself clear?

And then as a cow recedes into the distance:

> Keep your hands off my rear.

PETA limits its ethical exhortations to this one heartfelt issue. It eschews prudishness and Calvinist doing-without; in PETA's moral universe, ascetics are no closer to heaven than bon vivants. You like your little luxuries? Not to worry, just make sure animal skin isn't one of them. You insist on being fashionable? Fine, but not with leather and fur. Dark sexual fantasies to live out? You don't need black leather for that, either. "Dominatrices Whip Up Support for Fake Leather" led off one PETA ad campaign featuring demonstrators garbed "in sexy, 'sinthetic' outfits and bearing signs that read, 'Latex—In, Leather—Out.'" Wear "pleather," generic for nonleather alternatives. Latex, polyurethane—take your pick, so long as it's not leather.

PETA, it's been said, aims to "coax and sloganize leather into a state of political incorrectness"—until it's no longer the model to which any nonleather alternative must compare itself but instead a social malignancy, unworthy of emulation.

PETA's is not the only standpoint from which familiar hierarchies of real and faux are upended or claims like "100% Genuine Leather!" are seen as suspect or irrelevant. We saw earlier, for example, that, however deep its roots in nature, leather cannot be classed, even in principle, as entirely natural. And that however "chemical" its imitators, so is leather. In this chapter we'll confront other such paradoxes and provocations.

All through this book we've watched this one natural material, leather, fight off synthetic challengers, sometimes witnessing the struggle through its eyes, sometimes through those of its imitators. Of course, leather can be seen as just another material, like tinplate, cardboard, nylon, or beach sand—lifeless and inanimate. To imagine it fighting off competitors can seem the worst sort of conceit, like a Disney cartoon where broomsticks or teacups shudder to life, all shining eyes and gleaming teeth. But is it foreign to us as humans to invest inanimate objects with life and meaning? Chalices and scrolls are esteemed as sacred. "Diamonds are a girl's best friend." We cherish books, colored marbles, fine china, burls, baseball cards. So as we prepare to leave behind this enigmatic material, can we perhaps muster for it some little sympathy or affection?

Or then again, maybe not, not if by now our affection has been stolen by the Leatherettes, Corfams, and Amarettas of the world.

Or maybe it doesn't terribly matter: On the final page of *The Magic Mountain,* his long novel set on the eve of World War I, Thomas Mann bids farewell to his protagonist, Hans Castorp, affecting indifference to his fate on the battlefields of Europe. Likewise for us, if in a slighter way, as we part company from leather and the other materials whose company we've shared, and take one last lingering look back at the landscape of REAL and FAUX they inhabit.

ii. Faux Manifesto

The story of the quest for artificial leather is a story of claims. Proud claims. Careful, qualified claims. Fraudulent claims. Claims in patent applications and advertising. Implicit claims, and claims laid out straightforwardly, for anyone to judge or test. Claims that this time, finally, it's as good as the real thing. Claims in whispered assurances: *Psst, this stuff's better'n leather.* And it's been going on like this for the best part of two centuries.

In 1830 a German gentleman, a Herr Kummer, proposed combining paper waste with oils and varnishes, adding iron filings or fine sand, and squeezing the mixture, making for a species of artificial leather. "It was claimed," by one account, that shoes made from it were resistant to water "and as strong as real leather shoes."

In 1918, a time of wartime scarcity, Rudolph Obrist-Doos, of Lucerne, Switzerland, proposed the use of animal bladders that, while still raw, could be stretched and dried. The result, he claimed, was waterproof, flexible, durable, and "fully equal to leather." More recently, the embossed vinyl band of a low-end lady's watch was billed as reminiscent of bridle trails and hayrides and "nearly a twin to tooled leather." And practically yesterday, Kuraray called its microfiber-based Parcassio "as soft and supple as natural leather."

Soft as leather.

Looks like leather.

Feels like leather.

A twin to leather.

Better than leather.

Sometimes, faux sample in hand, we may be moved to reluctant appreciation. But more contemptible specimens are apt to leave us feeling we've fallen in with confidence men or snake oil salesmen, make us ask, *Who do they think they're kidding?* A typical account of a method used to make artificial leather boasted that the final product "defies detection": almost invariably, then, whether it be imitation leather, artificial, synthetic, or fake, a taint of deception clings to it.

Seen another way, however, the impulse to imitate yields a social good. New materials bring us the things of the world at lower cost or in greater quantity. And they challenge our snobbery, our smug assurance that we know what's good, better, best. Not surprisingly, they're routinely opposed by older, more entrenched competitors.

In his 1969 history of polyvinyl chloride, Morris Kaufman sought to account for resistance to plastics in general. "There is nothing like wood," ran ads in the London subway years ago, he reported. Their object was to beat back materials challenging wood, and they appealed to what Kaufman termed our "innate conservatism" about new materials:

> Older people well remember—and connoisseurs of oriental rugs all learn—the superiority of vegetable dyes over the synthetic variety. The old craftsman looked back with regret to handmade steel and every bibliophile knows that rag-paper is preferable to wood-pulp. When a new material seems to compete not only with traditional ones but with materials specially selected by the customers for their aesthetic qualities, "feel," and personal satisfaction, it inevitably encounters a subjective resistance which is very deeply based.

The innovation faces hostility that can be "intense and persistent."

You see it especially among those who have spent their lives with the old materials. A bookbinder trying Leatherette in the 1870s reported he'd "had a little to put up with from the obstinacy of the workman, who . . . often thinks 'there is nothing like leather,' and, if his prejudices to an *imitation* can help it, determines there never shall be." Likewise, when Giorgio Savini spent afternoons experimenting with the future Lorica, he faced mistrust and incomprehension from old tannery hands. Corfam, of course, stirred much enmity. "There is not a substitute now in the market, or being groomed for introduction, which does not seek to imitate leather in every possible way," the Tanners Council admonished in 1963. "They do not, and make no effort to, stand on their own legs."

But despite his decades-long roots in the industry, Mieth ("Slim") Maeser, the leather expert we met earlier, thought more broadly than that. In a paper prepared for delivery to Delaware Valley tanners, he asked: "What is wrong with frankly trying to imitate the good qualities of an established material in the production of a new one, particularly if at the same time an attempt is made to overcome some of the faults inherent in the old material?" Without that spirit of challenge, "there would have been no material progress, and we would be much poorer than we are."

Victorian-era artisans and inventors, as we've seen, made imitation an art form, yielding Leatherette, Fabrikoid, and countless other faux leathers. Sawdust and animal blood were mixed, heated, dried, and pressed into molds to make faux ebony. Beechwood was lathe-turned into chair parts, then painted and flecked, to resemble bamboo. Bronze was varnished, painted, or glazed to give it an instant patina, more normally the result of patient exposure to the weather. Is there not an elemental vigor even in the listing of them?

Scientists today mine nature's evolutionary experience for insights into new materials. They describe a bellowslike mechanism in cork, tell how ribbed iris stems resist bending. A classic example is Velcro, modeled on the burrs that stick to your clothes in the woods. And of course there's Ultrasuede, inspired in part by Miyoshi Okamoto's study of natural materials. A vital creativity, then, often lies behind imitation, right alongside its sleazy streak. "Good imitators," historian of technology David S. Landes has written, "make good innovators."

From this perspective, then—let's say it straight out—leather can seem downright stodgy. Bathed in the right promotional light, it can seem sexy, stylish, rebellious, or cool. But if you *really* wanted to rebel, you wouldn't turn to leather, so familiar and safe. Like other traditional materials, it serves as one strong strand in a safety net that keeps us clear of that abyss of imitation bamboo, artificial cork, and electronic bugles fitted with memory chips that play "Taps" at military funerals. *There's nothing like leather.* And that inclination to declare the familiar good enough, better than the barbarous new, makes leather profoundly conservative, an embodiment of the status quo. Besieged by edgy and indeterminate new materials hatched in labs last week, consumers seek comforting labels that promise Genuine Leather, Cuir Vrai, or Vera Pelle, finding in them an affirmation of all that's old, familiar, and true.

Leather, then, represents the entrenched order—and benefits accordingly. Some years ago two Korean researchers asked more than 400 consumers what they liked or disliked about various automotive seating materials, including cloth, pile, leather, and synthetic leather. Leather was broadly preferred; it was thought "luxurious," boasted "soft texture," offered "good seating comfort." But when the researchers actually tested leather against other materials, they ranked it low in its ability to transmit water vapor; it was apt to feel sticky on a hot day. Leather wasn't all that comfortable; yet people ignored the evidence of their senses and applauded its seating comfort.

The discrepancy bothered the researchers enough that they endeavored to explain. Natural leather, they pointed out, was porous. But the leather actually used in most car seats was (as we'll see) heavily pigmented and finished, clogging its pores; in practice, it was no better than plastic. Study respondents were asserting qualities traditionally imputed to leather that they didn't actually experience. It's as if leather at its best, the way it was enjoyed in the finest upholstery, was passing its cachet to poorer, overfinished cousins that bore the name leather but were anything but "natural."

But then again, what's so great about NATURAL, anyway?

Perhaps nowhere can we find a more contrary notion than in *À Rebours*, an 1884 novel by author and art critic Joris-Karl Huysmans, variously translated as *Against the Grain* or *Against Nature.* Huysmans writes that his aristocratic protagonist and narrator Duc Des Esseintes considered artifice "to be the distinctive mark of human genius." Of nature, he says:

> there is not a single one of her inventions, deemed so subtle and sublime, that human ingenuity cannot manufacture; no moonlit Forest of Fontainebleau that cannot be reproduced by stage scenery under floodlighting; no cascade that cannot be imitated to perfection by hydraulic engineering; no rock that papier-mâché cannot counterfeit; no flower that carefully chosen taffeta and delicately coloured paper cannot match.

The bored duke flees Parisian life and creates a succession of artificial worlds. He fits his house like a ship's cabin, flanked by a great aquarium that can produce "at will the various tints, green or grey, opaline or silvery, which real rivers take on according to the colour of the sky." The novel's peculiar conceit, then, is to disparage the work of nature and celebrate instead every human imitation or representation of it; in this it can seem a foretaste of today's virtual reality.

The habit of mind that deems nature's workings superior to those of men and women goes back at least to Aristotle. "If one way be better than another," he wrote, "that you may be sure is nature's way." Ever since, literature and philosophy have esteemed nature as teacher and model, human art and artifice standing low by comparison.

But while Des Esseintes's is an extreme, not to say pathological, sensibility at work, over the centuries others, too, have stepped up to rank human works as the equal of nature. "All praise of Civilization, or Art, or Contrivance," wrote John Stuart Mill in an 1854 essay, "is so much dispraise of Nature; an admission of imperfection, which it is man's business, and merit, to be always endeavouring to correct or mitigate." Another titan of Victorian thought, T. H. Huxley, expressed kindred ideas. Consider, he said, the garden, where flowers and plants are protected, through human intervention, from natural hardship; for all its luxuriant green, and however much it might stir the human heart, a garden is surely "artificial," no less than the flint tool or the cathedral. By these currents of thought alive in Victorian England, at least, human work does not stand below that of nature's.

So why, even in principle, place "natural" leather, say, on higher ground than Du Pont's, or Toray's, copy of it?

The Culture of the Copy is Hillel Schwartz's exhaustive 1996 study of imitation, reproduction, copies, and fakes in every area of life; it bears the subtitle *Striking Likenesses, Unreasonable Facsimiles*, chapter headings like "Second Nature" and "Seeing Double." Schwartz tells of ancient Egyptians who glazed plain ceramic beads to pass for turquoise and of highwaymen in search of booty deceived by polished spinel, its natural crystals resembling cut diamonds. Imitation, he suggests, is no aberrant departure from the true and the good, but is worthy in its own right. He writes, too, of how, with "foil-backed rock-crystal 'rhinestones'" from Bohemia and ceramic copies of antique cameos, "deceptive imitations came within reach of the middle class" and adds: "A single piece of well-worked costume jewelry was a treasure to a poor working girl."

What to make of this last statement? Contempt for cheap jewelry, made from imitative, second-rate materials? Or pity at the inferior taste of the working class, too easily satisfied? Or satisfaction that, thanks to the low-cost copy, they had "treasures" to enjoy?

As Jeffrey Meikle observed in *American Plastic*, plastic appealed in part "because it allowed familiar desires to be fulfilled more easily and cheaply through substitution." Take amber, the red-brown fossil resin of a now extinct species of pine tree. Beautiful, yes. But too expensive for any but small objects, like beads, certainly none so large as hairbrush handles or combs. But then along came the early celluloid plastics, which could be made to approximate real amber's transparency and color. Amber, Du Pont pointed out in the 1920s, could never be found "in sufficient abundance to permit its use in the manufacture of toiletware." But the company's synthetic substitute "brings this toiletware within the reach of all discriminating people."

Likewise for imitation leather. Ultrasuede isn't cheap, and neither is Lorica, but these are the exceptions. In the end, Du Pont dropped its Corfam prices. Leatherette went for as little as one-eighth the price of the real thing. A sofa in faux suede sets you back less than the same one in leather. Most synthetic leathers appealed because they made goods affordable that otherwise were not; consumers got *something like* what they hadn't at all before. Naugahyde's "nostalgic recollection of leather," writes Thomas Hine in *Populuxe*, "allowed the manufacturers of family-room furniture to democratize the great overstuffed chairs and sofas hitherto found mostly in gentleman's clubs and private libraries." A 1911 Du Pont brochure touted Fabrikoid for evening slippers:

> The costliness of colored leather and satin evening slippers, and the impossibility of keeping them clean enough to be presentable for more than a very few wearings, operates to place delicate evening slippers beyond the means of the average woman.

But low-cost Fabrikoid leather brought them within her reach.

Nineteenth-century French economic historian Georges d'Avenel heartily welcomed "the mass of cheap imitations flooding the marketplace," writes Rosalind Williams in *Dream Worlds: Mass Consumption in*

Late Nineteenth-Century France. "Instead of living with frustration, the humble could now enjoy the pleasures of being rich." D'Avenel reveled in how French men and women could eat off porcelain and enjoy factory-made rugs and wallpapers. True, mass-produced silk at 2 francs per meter in Parisian department stores wasn't the equal of fine Lyonnais silks at 600. But, wrote d'Avenel, "they make more people happy."

Cheap imitation? Cheap doesn't just mean "shoddy" or "inferior"; it means affordable; through rayon and other artificial materials, asserted Pauline G. Beery in *Stuff,* published in 1930, the chemist had "done more than any other single agency toward making a democracy of all the peoples of the world." If imitation leathers haven't always matched leather at its best, then maybe they've left *more people happy.*

And maybe Nature happier, too. An early proponent of celluloid argued that it gave "the elephant, the tortoise, and the coral insect a respite in their native haunts." No need "to ransack the earth in pursuit of substances which are constantly growing scarcer." Artificial substitutes eased pressure on natural rubber and pearls. *Barron's* in 1927 pictured Fabrikoid as indispensable since "all the cows on earth" could not satisfy the demand for leather for car and furniture upholstery.

Likewise for Corfam, which Du Pont wrapped in a flag of enlightened social consciousness: The population explosion and a globally rising standard of living meant more people wanted more shoes. Demand for leather, according to company projections, would outstrip supply by 30 percent by 1983. Thus, the new material's ethical virtue: "Not the least of our satisfaction is derived from a humanitarian aspect of the venture," a Du Pont official said at the time: "'Corfam' can make the difference between 'having' or 'doing without.'" No leather shortage materialized. But greater use of synthetics in the years since may have helped stave off scarcity. Without them, Max Kirstein, founder of Irving Tanning, said at the height of the Corfam scare, "hide prices today would be 50 per cent higher than they are now."

War breeds scarcity, too. And scarcity breeds *ersatz.* That's the German word for "replacement," or substitute, usually an unappealing one. We're stuck with this poor stuff, says *ersatz* coffee or *ersatz* leather, because we can't get the real thing, which everyone *knows* is better. By one definition, *ersatz* means "a cheap or inferior copy that can fool no one," typically spurred by economic hard times or war.

In Germany during the 1930s it was preparation for war that did it. The Third Reich's shoe industry, which depended on imported hides and was gearing up to equip the *Wehrmacht*, began testing *ersatz* leather soles, recruiting workers to walk in shoes made from them. This took a sinister turn during the war, the University of Göttingen's Anne Sudrow has shown, when a *Schuhläuferkommando,* or shoe-walking brigade, was set up at the Sachsenhausen concentration camp. More than 150 inmates were ordered to walk, under conditions amounting to a death sentence, all to test *ersatz* leather soles.

"Every loyal American should help save leather," read a Fabrikoid ad appearing four months after America entered World War I. No assertions of luxury, comfort, or resemblance to leather, just raw exigency:

> Uncle Sam is pointing the way. He is using leather substitutes for upholstery on all his Trucks, Ambulances, Air-planes and Ships. Will you help him?
>
> Whatever your business, make it your business to save leather. Every hide replaced with a good substitute helps furnish shoes for our armies, harness for our farms, belting for our factories—it helps win the war.
>
> What Leather Substitute Will You Use?
>
> Uncle Sam's Choice Is
>
> Du Pont Fabrikoid

And there, in the corner of the ad, discrete but inescapable, was Fabrikoid's trusty slogan:

> How Many Hides Has a Cow?

iii. Riddling Unions

Fake leather for defenseless animals, for the Fatherland, for Uncle Sam. Heavy going is it? But things fake can bear their own peculiar charm, too.

Earlier, in an Edith Wharton novel, we met a character put off by the "old-fashioned wood fire" of a home she visited and who missed a more fashionably fake one. In the years since, fake logs have been made from sawdust and paraffin, from almond shells, cardboard, and peach pits, even recycled coffee grounds. It's easy to lift an eyebrow at such insult to the sacred hearth. Or, for that matter, at aluminum Christmas trees or plastic flamingos in the front yard. But is there something charmingly dissolute about them, too? Like, maybe *enough*, finally, of all that leather, cashmere, mahogany, and gold?

That, anyway, is how Richard Artschwager came to feel. Artschwager was an experienced maker of high-end furniture who, as reported in Britain's *Spectator* in 2001, "became tired of nice materials." He was sick of the acres of fine wood in his shop. "So I got hold of a scrap of Formica," in "bleached walnut" no less. Formica was "the great ugly material, horror of the age," but he felt drawn to it and, in his early 40s, began making art with it. At least to some critics, here was evidence that Artschwager was no mere craftsman but a true artist. The "best" materials? No—too familiar, too Establishment. In these cheap, wretched materials, on the other hand, could be sensed a fresh, more artistically valid spirit. Plastic fakes bad? Or, rather, *ba-a-a-d*, and so funky, hip, innovative, and cool?

It's such insouciance that fashion historian Grace Jeffers enjoys in faux, in particular those fanciful, superficial qualities that lovers of top-grain leather, for example, might abhor. To Jeffers, whom we earlier heard sing the joys of Naugahyde, fans of the artificial are "sensual, playful, in-the-moment, with a great sense of humor." Turning the fetish for the natural on its head, they embrace the plebian pleasures of the faux. To them a Uniroyal exhibit in 1970 that featured hand-hewn beams and worm-eaten paneling, all made from polystyrene foam, might stir smiles, not grimaces. Even David Pye, whose *Nature and Art of Workmanship* otherwise lauds natural materials, could enjoy the fun: "No one supposes," he writes, "that the lady with pearls as big as birds' eggs round her pretty neck is flaunting the wealth of the Indies, nor would the imitation marble in St. Peter's in Rome deceive a child. . . . These things are open and rather cheerful bravura, not deception."

Sam Lange remembers well his days at Fabrikoid in the 1950s. The Newburgh plant made pyroxilin-coated imitation leathers in one building, vinyl in another, and had an unquenchable thirst for new, distinctive leather grains for the embossing rolls; no generic, plain-Jane patterns would do. So Lange, a young engineer a few years out of Rensselaer Poly, was regularly sent down to New York City to find them. He'd wander in and out of shops, storefronts, and lofts in the Seventh Avenue garment district specializing in hides, buy all the interesting ones, and haul them back to Newburgh. "We were shameless," he says. "We'd copy all the leatherlike looks we could find." Fakery? Stealing poor ol' leather's grain?!? *Come on.*

There, in New York City, on the prowl for leathery perfection, 20-something Sam Lange was having a ball.

These days, rock bands are lauded for copying old styles. "No longer," by one account, "must anyone apologize for being derivative. . . . Allegiance to the faded sounds of yesteryear is the new authenticity." Sidewalk purveyors of fake Louis Vuitton handbags get respectful *New York Times* treatment. A vendor reduces his $200 price to $180 but gets turned down. "But these," he's quoted as saying, "are the *real* copies!" In our culture today a perverse charm often attaches to the imitative and inauthentic.

Nineteenth-century critics of imitation, writes Jeffrey Meikle, understood its allure. Ruskin called imitation one of five sources of pleasure in art. "Whenever anything looks like what it is not, the resemblance being so great as *nearly* to deceive," he wrote, "we feel a kind of pleasurable surprise, an agreeable excitement of mind"; the con is *pulled off* and *seen through* at the same time. Purveyors of imitation leather beseech customers to compare their wares with the real thing: *Can't tell it from the original.* Are they themselves, one wonders, swept up in the satisfaction Ruskin identified? We ooh and ahh at the magician's sleight-of-hand even as she mystifies us. We delight in the Leonardo de Caprio character in *Catch Me If You Can* who passes for lawyer, doctor, and airline pilot. We're left stricken in wonder as the fake snubs its nose at the oh-too-earnestly real, all ethical second-guessing swept away by our pleasure in the deception.

Consider a 1988 exhibition in New York, carrying the imprimatur of the Museum of American Folk Art, called "April Fool: Folk Art Fakes and Forgeries." Real and fraud were exhibited in pairs, encouraging museum visitors to look with fresh eyes; given the $319,000 a goose decoy had recently fetched, one couldn't be too careful. Forgery a crime? Yes, but the better the fake, the accompanying catalog seemed to say, the more slack-jawed our admiration for the forger's audacity and skill. Of a group of fraudulent samplers, replete with stains and dirt, one dealer said: "These are very good fakes, and a lot of us who should know better have been fooled, sometimes more than once. The faker is imaginative." Said another: "Fakes are not all bad. They add a certain spice to the quest for the antique. Collecting would be 'dull sport' if everything were as it seemed." Here was a corner of the art world inhabited by contemptible cheats—or were they colorful rogues?

The phrase "technological sublime" has been used to describe the awe inspired by the likes of supersonic planes and great bridges. How about a "could've-fooled-me sublime" inspired by objects, people, or materials that make us think they are other than what they are? In *Promethean Ambitions: Alchemy and the Quest to Perfect Nature*, Stanford University historian of science William R. Newman writes of how the Greeks felt about art's illusionistic power:

> On the one hand, they display an awe at the artist's mimetic skill, while on the other they are clearly meant to mock the victim of the deception. Greek art delighted in the ambiguous tension established between these two poles. The skill that could rival the gods in re-creating nature was also the trickery that fooled the eye.

Viewers of Myron's famous bronze cow were captivated, writes classicist Deborah Tarn Steiner, by the "riddling union between the breathing body and the unmistakable fact of the inanimate bronze or stone."

The more perfect the likeness, the more teasing the riddle. The diamond necklace on some Hollywood starlet's neck at Oscar time: Real or fake? Identical twins: Which is Sally, which Sue? How sweet the triumph when the riddle is resolved, the deception laid bare. In 1885 a shoe trade journal told of "a certain make of cheap pump-sole brogans" whose artificial soles, made from ordinary pasteboard and covered with paper colored to look like leather, had fooled everyone. Finally, "a buyer noted for his keenness" saw through them, forcing prices down to a level "that comported in some degree with their worthlessness." Artful deception had crumbled before superior insight. I remember the satisfaction I felt, years ago in college, when I could distinguish my slide rule, eyes shut, from the identical model of a classmate's; slipping the mahogany slide back and forth, *the rough spots were different.* Twinned objects may seem indistinguishable to lesser mortals but not to those of us of refined taste or discernment. *We*, of course, can tell.

But what if we *can't* tell? What if the imitation is too good? Or, alternatively, what if we're not discriminating enough to tell the difference? In the 1940s, pioneer English computer scientist Alan Turing wondered how you could tell whether a machine could think. Well, he suggested, what if, sitting at a keyboard, you interacted with both a person and a computer, neither of whom you could see, asking nonscripted questions, getting answers, conversing. If at the end of an hour or so you weren't sure which was which, then you'd say the machine showed intelligence. Turing called it "The Imitation Game"; it comes down to us today as "the Turing Test," a landmark in the field of artificial intelligence. In 1994 a designer known for his fresh insights about crafts, Peter Dormer, extended the idea. Struck by machines that seemed able to mimic handcrafts, right down to "randomness, accidental quirks and less than perfect condition," he proposed his own version of the Turing Test: "If you cannot tell whether a piece of machined textile is hand-done or machined, then either the much-vaunted poetry of the handcraft aesthetic is a myth, or the same poetic aesthetic claimed for handcraft is also achievable through technology."

Maybe it's time, then, for a "Turing Test for Fake Leather": A faux leather that can't be distinguished from real leather would pass. And, like handcraft-mimicking technology, it would threaten us. For doesn't it seem as if we should *always* be able to tell leather from its imitators? Haven't we, somehow, an almost spiritual stake in the belief that, if nothing else, we should be able to distinguish something bearing the mark of Nature from the irredeemably inanimate?

But in fact we often cannot.

While researching this book, I was incessantly rubbing, pinching, and scrutinizing anything that resembled leather. Take me to a Pottery Barn, a hotel lobby, a doctor's waiting room, a leather goods store, or a friend's living room, and I'd soon be running my hands over this material or that, wondering: *Real or faux?* Sometimes, I'd pull out a magnifying glass, squinting for hair cells or evidence of embossing, occasionally turning a chair upside down, or a restaurant menu inside out, looking for bits of cloth, frayed edges, or other stigmata of man-made origins. Encountering samples whose leathery pedigree left me doubtful, at least without ripping them apart, my ignorance galled me. Galled me, that is, until I met leather experts, real pros, with years in the tannery trade, who sometimes couldn't tell either—and who seemed content, absent lab tests or microscope, to say so.

Tottering on that knife's edge of uncertainty, of two minds as to whether a wallet comes from a cow's hide or a petroleum pipeline, can leave you flustered. I saw this among friends I recruited for a little experiment. Having gathered numerous samples of real and imitation leather, I mounted them each on cards and asked my subjects to tell—first at a glance, then at a touch, then with a squeeze—which was which. Many were fooled by a piece of dried wet-blue that felt like cardboard or by one of Lorica flesh side up. Their verdicts often changed at each stage of the experiment; sometimes sight deceived, sometimes touch, sometimes both. But once the truth was out—faux revealed as faux, real as real—you'd hear squeals of wonderment from them at the deception: *Oh, I would never have guessed.* Or nervous chatter about why they'd got it wrong. Or a gasp of admiration for just *how* like leather it felt. The obvious ones, real or faux, inspired none of this. But when they had a piece of it right there, in their hands, and *still* couldn't tell for sure? Then, that "riddling" uncertainty, that suspense of not knowing, was like an erotic itch demanding satisfaction.

Among thinkers drawn to the borderland of the faux, none has been more influential than French postmodernist philosopher Jean Baudrillard, whose *Simulacra and Simulation* explores the muddled juncture between real and fake. Writing in an epigrammatic style riven with paradox and nuance, Baudrillard distinguishes, for example, between "pretending" and "simulating." The first, he writes, "leaves the principle of reality intact," the line between real and faux sharp. The second, simulation, "threatens the difference between 'true' and 'false,' the 'real' and the 'imaginary.'" Kuraray made just such a distinction in the 1964 patent we visited earlier: Whereas once there'd been just "imitation leather," Kuraray claimed its latest and best was truly "man-made leather"; in Baudrillard's language, Kuraray no longer aimed merely to "pretend" but to "simulate," erasing the edge between original and simulacrum. Indeed, for at least 40 years, back to Corfam, creators of new synthetic leathers have routinely offered micrographic cross-sections of their products, set beside that of leather, aimed at making something close to this higher claim—true simulation.

And when the boundary is breached and we no longer know what's real and what's not? Then, with real "no longer what it was," writes Baudrillard, "nostalgia assumes its full meaning." In the Levenger catalog we peeked at before, we are meant to respond to its backward-looking leather desk blotters and briefcases with a kind of nostalgic ache, just as we do its retro fountain pens and old-world globes. And would a book like this one, for example, be written or published in a less nostalgia-tinged age, when leather was still just the workaday stuff of boots, bridles, and industrial belts?

Baudrillard's own choice example is the hologram, that laser-generated, seemingly three-dimensional re-creation of an object in empty space that stuns science museum visitors with its verisimilitude. "Holographic reproduction, like all fantasies of the exact synthesis or resurrection of the real," he writes, is no longer real but "hyperreal." The hyperreal occupies what Baudrillard calls "the other side of truth," in some sense eradicating the original. "This is perhaps why twins were deified, and sacrificed, in a more savage culture; hypersimilitude was equivalent to the murder of the original." What, then, of some faux leather of the future, some new material wonder yet closer to leather than any seen today? For Baudrillard this exemplar of the "hyperreal" might make leather from animals seem alien and unreal.

In *The Culture of the Copy*, Hillel Schwartz wrote that only in a culture of incessant imitation "do we assign such motive force to the Original"; set beside its imitators, it is transformed, granted larger significance. The original of anything "speaks to us in an unmediated way, an experience we seem to believe we have lost between ourselves, human to human." Pressed by ever more persistent imitators, it can seem that leather bears a new and heavier burden of meaning on its poor shoulders.

iv. Mock Croc

Not that we can be so sure these days just what is so "original" about leather, or so "real," or so "natural."

Before World War I, Du Pont boasted that the grains it embossed on Fabrikoid came from hides selected as paragons of leathery perfection. Well, similar winnowing takes place every day, around the world, in the making of leather itself. By some estimates, 80 or 90 percent of the leather you see has had its original surface stripped away, then been coated and embossed, just like Fabrikoid; a handbag or shoe bearing its original leather grain is the expensive exception. Think of the fresh-faced 19-year-old supermodel, her beauty so natural, her complexion so perfect, that powder and rouge only sabotage them; likewise the best leathers, which need only a little oil and dye to be shown at their best. But the others? Leather may cling to the mantle of "natural," but more typically it's processed to within an inch of its life, overlaid with unnatural veneers.

England in 1905 saw publication of a report, *Leather for Libraries*, that lamented the lower quality of leathers used for bookbinding. Half a century earlier, it recorded, the age-old practice of decorating leather by mechanical impression had taken a bad turn when, through electroplating, it became possible to reproduce the grain of expensive skins on "sheepskin or other inferior leather." In the years since, sheepskin had virtually disappeared as such, "only to reappear as imitation morocco, pigskin, or other high-priced leather," the deception distinguishable only under a microscope.

Long before Leatherette, Naugahyde, and Corfam, in other words, embossing techniques like those used to make imitation leather were used to make real leather what it wasn't; those big Standex embossing rolls, for example, impress not only vinyl and polyurethane but real leather. In one common practice, leather whose top grain is deemed too blemished to use as is is buffed away; topped over with silicone, nitrocellulose, or other coating; and embossed. "Corrected grain," it's called. The original grain, with tick bites, scratches, and wounds the animal suffered in life, yields to a choice, idealized one—in short, a facelift. Talk to American tanners and they'll blame feedlot operators for the poor quality of raw hides; those reaching the tanner bear ever more barbed-wire holes, insect bites, and disfiguring crusts of manure—which must, of course, be "corrected."

Modern life is "no longer a question of imitation, nor duplication, nor even parody," writes Baudrillard in one of his choicer epigrams but "of substituting the signs of the real for the real." True enough, plainly, for Naugahyde. But perhaps yet *more* true for an aesthetically imperfect cowhide, its grain layer buffed away, impressed with a different grain, from another cow, from across the feedlot or across the world.

And when it's no cow at all, but this time the grain of some other species entirely? A common example is crocodile. Cowhide embossed with the variegated, platelike mosaic of crocodile hide is leather, yes. The face it presents to the world, however, is not that of the original animal but quite another of God's creatures. The result, if you think about it too long, can seem as unsettling as Frankenstein's monster or a face transplant. Not to mention deceptive. The Federal Trade Commission insists that such species transformations be revealed to the consumer; "Mock Croc," you'll hear it called.

May we venture further into this strange, unnerving back country of appearances and deceptions? Consider a desk chair I bought not long ago, consisting of a metal frame covered with "pigment-dyed leather." It looked good in its deep chocolate brown, was comfortable, and not too expensive. But you'd never actually *touch* this leather, doubtless a cheap split, unless you fished down beneath the seat, as I did, to pry it up. For it was laminated to a coating of rubbery plastic, and it was *this* upon which you actually planted your behind; this was no light "finish" barely obscuring the leather beneath but a coating as thick as a picture postcard, its inner foam layer surmounted with a hard outer crust—lightly embossed, as might be expected, with a leather grain. It was like going to a costume ball and meeting someone wearing a Muhammad Ali mask—which then, at evening's end, was peeled off to reveal an older, less formidable . . . Muhammad Ali.

An importer of Italian furniture trumpets the "High Protection Leather" of its upholstery. By this it means hides that

> undergo careful processing to make them highly resistant to liquids, sunlight, wear and tear. This treatment protects them from stains and minor abrasions while reducing everyday maintenance to an absolute minimum.

Its ads boast "100% Leather. No vinyl!" But what if that "careful processing" yields leather that feels like vinyl? Leather's most ardent champions lament just this—that it often feels less the way leather "should" feel. The little experiment I foisted on my friends spoke to the tactile and visual success of synthetic leathers but also to the failure of today's real, yet "unnatural," leathers to set themselves apart from plastic.

Consider staking, the seemingly straightforward process of pummeling leather with vibrating mechanical fingers to make it softer and more pliable. Diderot's *Encyclopédie* shows the special hammers used to soften leather in his day. Now it's done with machines, but it's the same idea. The subject came up on an online discussion forum for leather chemists. "Dear Fellow Tanners," it began, "What is your feeling on proper conditioning before staking?"

You need 20 to 22 percent moisture content, came a prompt reply; finished leather runs about 12 to 15 percent. Get it by vacuum drying the wet tanned hides, laying them out flat on a screen through which the moisture is sucked away, monitoring moisture content until you reach the right figure. Or else dry the hides completely and then *add* moisture to get your 20 percent. From here the discussion broadened to structural and nonstructural bound water, the hygroscopic point, and how acrylic resins soften leather. . . .

Some tanners, put in a new correspondent, staked leather while wet. That demanded special belts, "but very nice tight leather can be made with excellent footage." *Footage?* He meant that the leather, sold by the square foot, would actually gain in area. An earlier writer came back to wonder whether this wet staking might be used, deliberately, not just to soften but "to extend the area gain" achieved in earlier processing. "What a novel approach!" it struck him.

Yet right there, online, he had second thoughts:

> Then again, when do we reach the limit of stretching leather to gain the optimum area while sacrificing so many other physical attributes? Eventually, you have to have a leather with nearly no elasticity. That is, [you] set it out, stake it out, toggle it out, plate it out, etc.,

these referring to mechanical processes that increase footage.

> What do you have left? Leather?

And how can the culmination of this tricky, technology-abetted human intervention be termed "natural"?

Perhaps no word in recent years has been more utterly stripped of meaning. In the late 1970s the Roper public opinion company asked 2,000 adults to choose from a list the two or three things they associated with a "natural" way of life. Virtually every item on the list was tapped. A simpler lifestyle? That was natural. Avoiding artificial and processed foods? That was natural, too. Being yourself? That, too. Making things yourself? Yup. Wearing simpler clothes? Cutting down on consumption? All these and more fit under the great tent of NATURAL. The idea had insinuated itself so deeply into American consciousness, had become so fuzzy and vague, that it meant almost everything and, aside from being good, almost nothing. This maddening blur of un-meaning, of course, has settled on the labeling of natural and organic foods as well. And on leather itself, which is hardly the first natural material to make us ask what constitutes "natural" in the first place.

In *Promethean Ambitions,* William R. Newman took as his subject alchemy, the predominant proto-scientific enterprise of the late middle ages and early modern period. Roughly, alchemists sought to transform base materials into noble metals, especially gold, through a would-be "Philosopher's Stone." That, at least, is the standard summary, one

Newman enriches. Aristotle, he tells us, distinguished between *imitating* nature and *perfecting* it, a distinction maintained through the whole middle ages. Those holding out for this second and higher alchemy aimed, says Newman, for "the genuine conversion of commonplace materials into entirely distinct substances of much greater value." No imaginative leap is needed to apply the notion to synthetic leather: cheap, commonplace monomers yield substances all but equivalent to the more highly valued leather. But might it apply as well to leather itself?

We've seen how leather, for eons tanned with locally grown vegetable materials, gave way, in the late 19th century, to more sophisticated technology based on chromium salts and bearing every hallmark of a chemical process industry. We've seen, too, how tanners, by judicious use of beamhouse operations before tanning, and by fat liquoring, retanning, and modern finishing methods after tanning, can make the same hide come out as hard and unyielding as a fence post or as soft as a baby's bottom. Much in the way of smoke and mirrors, then, lies behind leathermaking; they don't call it the tanner's art for nothing. So can it be seen also as a species of alchemical transformation? A "commonplace" material, raw animal hide fresh from the abattoir, is converted into a "distinct substance of much greater value"—that is, leather.

Wöhler's synthesis of urea cut across the once seemingly sacrosanct border between human work and Nature's. There is, after all, but one Nature; leather and its imitators share it. Yet as Newman highlighted another scholar's point: "The distinction between the artificial and the natural did not disappear" that day in 1828 with Wöhler's discovery. Moreover, while it's one thing to assert that in some abstract or philosophical sense collagen is no less "chemical" than vinyl, leather no more "natural" than synthetics, it's quite another to blithely abandon one's everyday predisposition to the contrary. So if, for the moment, we cling to leather as "natural," might we be forgiven for asking why its nature is so routinely processed out of it?

Jean Tancous, the expert on leather defects we met in an earlier chapter, tells of a Cadillac she and her husband drove years ago. "I loved it for its seats," she remembers; they were plush, leathery, and luxurious. These days they have a new car. "It's got leather upholstery," she says, "but it feels artificial." So winter and summer alike, Tancous throws an old sheepskin over the seats; it feels much better to her than the leather.

In the competitive crucible of automotive upholstery, paradoxes of real and faux bubble up all the time. Americans, it was said at a recent meeting of the American Leather Chemists Association that highlighted this commercial niche, live in their cars. They spill Cokes over their upholstery, splatter ketchup from Big Macs. "In the last few decades, the automobile has changed from simple transport to a living environment on wheels," one speaker observed. Dress is more casual—and harder on upholstery. Once, driving a luxury car meant a business suit. Now it can mean greasy jeans. Car upholstery must stand up to more day-to-day abrasion and abuse. To Arizona heat. To Minnesota cold. It mustn't fog up the windshield. It mustn't smell objectionably; in one test a sealed container holding a sample is opened after two hours and sniffed by three testers who rank its odor from "imperceptible" through "perceptible but not disturbing" to "intolerable." So hides destined for leather upholstery are typically whisked along on conveyors under rapidly spinning spray guns that paint them with heavy protective pigments and finishes. "Automotive leather is a far cry from what consumers expect in fine clothes and accessories," noted one trade magazine article. "Meeting 10-year/100,000-mile durability standards frequently leads to leather that looks and feels little different from a good piece of vinyl."

Actually, much of what's called leather upholstery *is* vinyl; that is, you get leather "seating surfaces," all right, but the rest of the interior is done up in vinyl to match. "A" surfaces and "B" surfaces: That's the term of art in the business, says Jeff Post, formerly vice-president at SanduskyAthol, the vinyl manufacturer, more recently design manager for color and materials at Ford Motor Company. "A" surfaces are those your body touches, like the seat itself. "B" surfaces are everything else, including consoles and seat backs, even the sides of the seat cushion. Examine years-old upholstery, in particular the border region between "A" and "B," and on the leather side of the seam where they meet you'll likely find a dry riverbed of loose pigment flakes encrusting leather whose original surface has been stripped away. If you then sight across the line of stitching to the side of the seat cushion, you'll see wrinkles, ridges, and veins reminiscent of leather—only shinier with the years, the color no longer quite matching, plainly plastic.

The discrepancy in color and texture might not matter much in an old car, but in a new one the two surfaces had better disguise their differing pedigrees. Enter "color matching." At Naugahyde, recalls former director of styling Michael Copeland, they'd go into a color-matching meeting with a dozen or more samples each of leather and vinyl and spend the whole time fussing over shades of tan. At SanduskyAthol the factory has a special room in which to compare samples in various kinds of light—fluorescent, natural daylight, and setting sun. They'll take color readings from a sample, record departures from spec, and adjust the manufacturing process accordingly. "The vinyl suppliers have done a fine job replicating the grain and appearance of leather," a top designer told *Automotive Industries.* "I even have a hard time telling the difference myself"—which is, of course, just the idea.

Post's old company, he says, made one leathery vinyl, SUTTON, for high-end Chryslers, embossed to show off "natural" markings. One sample of it shows a deeply pebbled surface, hints of hair follicles showing through, and, oh . . . right there in the middle, its regular grain pattern interrupted, a blotch the diameter of a pencil eraser—a tick bite. A tick bite in vinyl. They were making it down in Mexico, reports Post, when a conscientious worker pointed out the defect to his boss. No defect, he was told, it was supposed to be that way.

A travesty, this intrusion of the synthetic into the realm of the natural, this color-matched, tick-bite-ridden vinyl-the-usurper locked in unseemly embrace with gentle leather? Of course, as we've seen, the embrace is mutual, leather often so plastered over, so unnatural, that it no longer feels like leather at all. Unrelenting cost pressures maintain the status quo. The low-quality hides needing all that surface doctoring are cheaper—$20 or so a square meter, a few years ago, compared to maybe $50 for good aniline leather. When, as one conference speaker boasted, a new post-tanning treatment can squeeze and stretch a 50 square foot wet-blue into 58 square feet of finished leather, how it feels at the end is apt to be forgotten; it's no longer, certainly, so sumptuous to the touch. An otherwise respectful review of one American car carped that its leather upholstery "looks cheap, like an old Naugahyde jacket."

But it's not only by look and feel that leather car upholstery veers this way, then that, across the faux-real divide. We are in Hazlet, New Jersey, just south of New York City, home of International Flavor and Fragrance, a company that sells $2 billion worth of the ingredients that make modern food and household products taste and smell as they do. On one side of Route 36 is the company's R&D center, where test subjects sit in booths sniffing and tasting sample meals. Or picking from swatches of velvet, leather, cotton, chenille, and other materials that they can touch but not see those that remind them of a particular fragrance. IFF's annual report shows two girls happily slurping ice cream bars; describes "cocoa-extender" technology that lets chocolate makers get by on less chocolate; tells of stroke victims introduced to familiar smells to speed their recovery. Here at IFF, Stephen Warrenburg speaks with pride of a computerized flavor thesaurus. Here, Guillermo Fernandez earns his living as fragrance development manager, or "Nose." He's worked on projects for Jaguar, Ford, Chrysler, and General Motors, and one thing he's done in his long aromatic career is help leather smell like leather.

"Dark and feral, noble and feline, soft and velvety—the smell of leather strikes various notes all closely associated with the skin," writes French writer Anne-Laure Quilleriet in *The Leather Book.* In a 1960s vintage leather shop it's the dark musky smell that seduces you. Buy a new wallet or purse and it's the aroma wafting up from it that may be why. With disagreeable tannery smells needing coverup, leather has a long history at perfume's side; France's perfume capital, Grasse, Quilleriet points out, started out as a tanning town. And perfumers many times turn to leather for inspiration. One fragrance created in 1996, Cuir Mauresque, was said to create "an olfactory impression that is more leathery than an actual leather."

How does "actual leather" smell? Mainly, said Waldo Kallenberger in an online forum for leather chemists, the "'traditional leather' smell" is fat liquors, fish oils, and sometimes vegetable extracts. These days you can buy it in a spray bottle and spritz your car with it. Coming soon are invisible polyurethane microcapsules that release eau-de-leather fragrance when crushed; sprayed on the flesh side of finished leather, they'd give up their aroma only gradually. As an account in *World Leather* had it, the "pleasant new smell that greets you when you first get into a new car with a leather interior will still be there" long after.

One forum correspondent said he didn't want people messing with real leather's natural smell, certainly not with anything fruity or floral: "It takes away the genuineness that is associated with leather. There would not be any distinction between synthetics and leather." Of course, maybe some tannery could survey the fragrances of veg-tanned leathers, select the best, have it manufactured by a perfumer, and spray *that* on finished leather. Hmm, well, there was a problem with that, too, he realized as soon as he wrote it. "The synthetic leather lobby would also start using it to promote their stuff as genuine."

Or maybe not. A smell-and-taste research foundation in Chicago reportedly found that people *preferred* the smell of artificial leather. Here was testimony to the inauthenticity of modern life, it seemed to Britain's *Guardian.* "So, in the US now," it noted, "leather seats in cars are impregnated with the smell of artificial leather."

IFF's Fernandez, a trim, angular-featured man who emigrated years ago from Castro's Cuba, won't say, of course, just what goes into his leather-bound scent. In his office in the company's creative center, plastic tubs each the size of a small kitchen trash bucket are filled with leather samples. Which, he asks, do I like best? I pull the lid off one and drop my head down for a whiff. Then, as Fernandez instructs, I draw back and sniff the top of my arm—presumably a neutral and consistent odor, the aromatic equivalent of the bite of bread taken between sips at a wine tasting. Then it's on to the next sample. Could I tell the difference between them? No, and even if I could I wouldn't be able to find words to describe it. But the Nose can; he nods toward one tub, assures me its fragrance is fresher and more authentic.

IFF doesn't normally charge customers for its research, not directly, anyway; rather, it sells the fragrances and flavors themselves, some 35,000 of them. A perfumer orders a new smell made up. A technician in the brown-bottle-dense lab makes up the recipe. A nose, like Fernandez, evaluates it for how well it's apt to satisfy the customer. The company has worked with vinyl makers, too, but Fernandez's assignment right now is for a Detroit car maker's leather interior.

There's not much you can do with leather, he admits. Just enough, though, that one luxury car maker could think to suggest the smell of cigars from a 1930s gangster movie. Another, an American car manufacturer, wanted nothing more exotic than the hint of a fine briefcase or new pair of shoes. *Something* distinctive. What nobody wants is that generic new-car smell, which Fernandez explains is just the lingering odor of vinyl and carpet adhesive.

Today, in a temperature- and humidity-controlled room, they're concluding tests to make sure they have it right. The size of an office cubicle, the glass-sided room is one of six, three on either side of a little corridor; you can purge its air in 30 minutes flat and try something else. In this one sits a large hide, tan and bland and overfinished, thrown over a chair, giving off its aromatic emanations, waiting for Fernandez or one of his colleagues to crank open the cubicle's little window, stick his head inside, take a whiff, and pronounce it just right.

Or else decide it needs a bit more tweaking to get that good, natural leather smell.

v. Material Matters

The leather that aficionados deem most desirably natural is that with an aniline finish, or aniline leather. Developed in the late 19th century, aniline dyes imbue leather with rich, transparent color; no top-heavy pigment, thank you, just the welcoming look and feel of leather at its best and barest, its original grain showing through. But in fact, many so-called aniline leathers aren't; while they boast a natural look, they are not, chemically speaking, aniline. In its mid-1990s review of labeling requirements for leather, the Federal Trade Commission heard Leather Industries of America assert that the term aniline "does not imply that the leather has been dyed with an aniline dye." To this reasoning, the FTC gave its assent.

We've seen leather as the overworked product of harsh chemical soups and mechanical pulling and pummeling; blemished leather "corrected" with idealized versions of itself; cowhide represented as crocodile; split leather made to pass for top grain. Just how deep does leather's impiety go? Consider the bewildering variety of names applied to leather down through the years. "Kidskin," one might suppose, should come from the skin of kid, or young goat; "chamois" from that of the chamois, the mountain antelope of Europe. In fact, Thelma Newman recounted in *Leather as Art and Craft*, both terms today usually refer to specially treated lambskin. "Morocco" originally meant sumac-tanned goatskin with a characteristic red cast. "Cordovan" was horsehide. "Russian leather" was calfskin dressed with birch oil. No more. The working craftsperson, wrote Newman, had to accept that "modern leather technology has advanced to the point where *imitations look like the names they take.*" The italics are hers.

Leather *dis*honest, *un*natural, *in*authentic?

During the mid-20th century, leather had no more thoughtful champion in England than John W. Waterer. The son of a printer and a schoolteacher, Waterer served an apprenticeship in his uncle's leather goods company, then after World War I became manager of its luggage department and its chief designer. During his 40s and 50s, he began expressing his views about leather, art, and design in articles, books, and radio broadcasts; some of them bore a moral or ethical tinge.

There were, said Waterer, three forms of dishonesty in design:

> It is dishonest for a material to masquerade as something other than what it is; as for . . . textile to imitate leather. It is dishonest for an article to be disguised as something else; for example, an electric lamp to simulate a wax candle. It is dishonest for an object to pretend to be made by a process totally different from that employed; for instance a plastic moulding to imitate hand-worked wood.

Waterer, who died in 1977, was writing at a time when vinyl was overtaking Fabrikoid-like faux leathers, and heavy pigmented finishes were corrupting leather, a trend he held in contempt:

> Superficially, modern finished leather may appear, to some, more attractive, with its perfectly even colour in bewildering variety, and with many of its natural features removed or hidden; but, in many cases, the development of these finishes has produced a strong tendency to artificiality.

This wouldn't do. Leather must remain true to itself, or else become "lifeless and uninteresting." Perfectly even color, for example, "causes it to resemble artificial machine-made products," discrediting leather, the good stalwart material he held up as paragon. Blurring the line between leather and, say, leathercloth was "unfortunate for both materials," he wrote: "Both have excellent and unique properties but of an entirely different character." No need to compete. "It is a species of dishonesty," Waterer once declared, "which causes one material to imitate another."

Dishonesty quite literally, according to the Federal Trade Commission. "Leather content representations likely are material to consumers," it asserted in explaining its revised leather-labeling guidelines; that is, it mattered to them. So, play fair:

> It is unfair or deceptive to use the unqualified term 'leather' or other unqualified terms suggestive of leather to describe industry products unless

> the industry product so described is composed in all substantial parts of leather.

Make nonleather look like leather and what it really is must be disclosed. Likewise, as we've seen, for leather "embossed, dyed, or otherwise processed so as to simulate the appearance of a different kind of leather."

Fabrikoid, *circa* 1915, might well have violated the FTC rule decreeing that you couldn't deceive, either, with symbols or trademarks, such as any "stamp, tag, label, card, or other device in the shape of a tanned hide or skin or in the shape of a silhouette of an animal."

But the commission left room for "honest" materials. Its guidelines triggered disclosure

> only when a product appears to be leather and is not. Many synthetics are intentionally made to simulate the look of leather, apparently because many consumers prefer leather. Other synthetic products, however, are clearly and visibly synthetic.

Visit Nassimi Corporation, the New York-based distributor of vinyl-coated fabrics, ease back into a high-backed swivel chair in its conference room, and you're enjoying upholstery that doesn't so much as hint of leather. "Vinyl and proud of it," says Ed Nassimi. *Imitation leather needn't imitate leather*, which is a silly enough way to suggest that fabrics coated with the nitrocellulose, vinyl, or polyurethane often going into faux leathers needn't look like leather itself. The original Leatherette introduced in the 1870s, for example, could be had with a variety of leather effects, but you could also get it with a "diced grain" whose diamond-knurled surface looked like the man-made thing it was. Likewise Naugahyde, which despite that "hyde" doesn't always mimic cowhide. Some Naugahydes imitate particular fabrics. Others look only like their eccentric selves. ZODIAC, in greens, golds, and hot pinks shot through with metallic flakes, looks like it just got off the plane from Vegas and resembles no leather Nature would ever claim as its own. With such materials the old hierarchy with leather at the top and its imitators at the bottom is not inverted—as PETA, say, might prefer—but scrapped altogether, with no top, no bottom, no look or texture better than any other, no material anything but what it is.

"My goal is to make people feel a connection with the man-made world we live in," says Grace Jeffers, the design historian. Jeffers exults in the plasticness of plastics, the Naugahydeness of Naugahyde. "I love sitting on it. It's weirdly warm and cold at the same time. I love the cleanability of it, all the colors. It feels like a shiny candy coating over the warm meatiness of the nuts." But designers have been too timid with it, she feels. Naugahyde "has not been pushed to its limits."

Subscribing to similar ideas is Jeff Post, whose whole career is steeped in vinyl. While still at SanduskyAthol, he concluded that however much he and other vinyl makers tried, vinyl would never look or feel like leather; most people saw it as simply "fake leather, something that sticks to your legs in summer," forever inferior. But it needn't be that way, he insisted. Determined to "redefine the visual brand identity of vinyl," he tried selling customers on nonleathery vinyl that was "not Naugahyde, not leatherette, not fake anything," but simply itself. A sometime actor with television and Hollywood credits, Post experimented with outrageous patterns and colors but regularly got shot down for his trouble. Think up something fresh and all his automotive customers could say was, "No, let's put SIERRA on it," referring to a generic leather grain. Once, he enlisted the help of a designer friend. Go wild with possibilities, he told her, any printing and embossing you like, but whatever you do, nothing that looks like leather. One of her minor masterpieces had a palm frond motif, all blacks and deep, moody greens, hauntingly beautiful. Why not? "The consumer I design for," Post says, "is very hip to how cool plastic can be."

That sensibility was more widespread after World War II. Then, "plastic seemed like a magical boon," write Robert Gottlieb and Frank Maresca in *A Certain Style*, their affectionate look back to an age when handbags made of Lucite were in fashion. Designers created fanciful sculptural shapes of solid plastic in tortoiseshell, butterscotch, and pearl, outlandishly ornamented, reminiscent of camel saddles, pagodas, and bowties, nothing subtle, testaments to plastic's infinite malleability.

Let plastic be plastic.

In an essay, "Speaking Volumes," in the airline magazine *Hemispheres*, Cynthia Reece McCaffety looked back fondly to the days of the old door-to-door encyclopedia salesman. He'd no sooner get in the door, she remembers, haul out Volume 1, and she and her parents were his. "I was weak with desire. . . . My fingers caressed the simulated leather bindings, and my breath was shallow and fast." Years later she unpacked the old volumes from the cartons in which they'd lain too long. "The edges are a bit worn, so the weave of the fabric shows through the simulated leather, tampering with the illusion of grandeur. But standing on the shelves, their gold lettering still gleaming, they have a singular and timeless beauty."

What are we to make of this story? That the old volumes had the same hold on her as when she was young, so that whether bound in real leather or faux didn't matter? Or that she was aware enough of the binding to remark on it, the discrepancy between it and real leather weakening its impact, so that it *did* matter?

It's possible to charge through life oblivious of one's material world, caught up in larger dramas of work, love, and family, blithely unaware of, for example, how books are bound. But in ways we sometimes scarcely notice, often at the very edge of perception itself, materials do reach out to us, contributing to our satisfaction and happiness or else detracting from them. In this book, it should be plain by now, I mean to say that yes, what things are made of does matter.

"Whole eras in the life of man are known by the materials upon which their technology rested, namely, the Stone Age, the Iron Age, the Bronze Age." So wrote psychologist Ernest Dichter, whose private research institute conducted some 2,500 studies of human motivation in the 1950s and 1960s, in the preface to his book *Handbook of Consumer Motivations.* We're well into the Synthetics Age now, of course. And, beyond that, a Virtual Age which, through pixels and digital imagery, draws us yet further from the leather, bamboo, silk, gold, and granite we've had our hands on for most of human history. We slip deeper into this epoch of unreality each time we touch a material that never before existed, or grasp something made of one material dressed up to look like another, or stare weakly at a computer screen bright with images of Baghdad mosques, origami, ebony, or John Lennon, that are all, every one of them, just so many identical phosphorescent pinpoints. "Faux wood, true practicality," speaks the headline for a line of kitchen blinds. Pillows of polyester squeeze out goose down. Traditional materials fight back. The Natural Stone Council laments "imitation engineered products." The National Christmas Tree Association battles aluminum and plastic rivals. A little company in northern Vermont, Island Pond Woodworkers, makes tables and chairs from "character wood," its streaks and knots more typically written off as defects.

"This is wood that reflects the forest," says Island Pond. "If you cut out all that, you might as well use Formica."

Such small unremarked breaches of the faux-real frontier lend to 21st-century life some of its distinctive flavor. For they're felt not symbolically or metaphorically but at the tips of our fingers and in the synapses of our collective visual cortex. "The story of synthetic substitutes for leather," a German footwear journal noted in 1976, demonstrates "mankind's age-old urge to find and exploit new materials." Each instance may be but a trifle, but these substitutions, emulations, and imitations, tens of thousands of them in a lifetime, together reach down into us. As they did wearers of Corfam. As they did that lover of encyclopedias, Ms. McCaffety.

As they did me, when I was 14 and my father, as he did every three or four years back then, brought home a new car. That year the car he parked in front of our house in Brooklyn was a convertible—a big, black 1961 Mercury with seats upholstered in acres of bright red leather. My dad? In a convertible with leather upholstery?

Cool.

But when I clambered inside and slid across those wide red bench seats, I knew they weren't leather. From a distance they were OK; you could see the dimpling in the surface, its irregular, seemingly organic variegation. But look just a little closer, touch it, and it had a kind of hard-edged, crinkled gloss that leather never had: Whatever was supposed to be leathery about it went down only so deep, about the thickness of a sheet of paper. Those seats weren't leather, and all you had to do was touch them to know. *Dad,* I whined, *why didn't you get the real thing?*

David Pye, that insightful commentator on materials and their working, all but denied that "quality" lies within material at all. "Material in the raw is nothing much," he wrote in *The Nature and Art of Workmanship.* "'Good material' is a myth." Fine English walnut? "Most of the tree is leafmold and firewood." Only *worked* material, he insisted, can have quality.

> We talk as though the material of itself conferred the quality. Only to name precious materials like marble, silver, ivory, ebony, is to evoke a picture of thrones and treasures. It does not evoke a picture of gray boulders on a dusty hill or logs of ebony as they really are—wet dirty lumps all shakes and splinters!

Or, for that matter, fine calfskin that starts off as unhaired hide, fresh from the killing floor.

But in the end, whether or not we can quite say why, the *stuff* in things matters. A visitor to Majorca, the island resort, encountered a sign that read "Welcome to Inca, Leather Town." "Rip-off Town" was more like it, he reported in an Internet posting. He'd bought a jacket in the market only to later find it was "made of imitation leather (PVC) despite the seller's insistence that it was 'top-quality lambskin.'" He didn't specifically complain about the appearance or quality of his purchase; but it was supposed to be leather, it wasn't, and that was quite enough to gall him.

A mail order house offered "a museum quality action figure" of General Custer at the time of his Last Stand, boasting of its realism. The toy featured "33 points of articulation," compared to only 21 for the GI Joe of the 1960s. Custer's rifle and pistol were die cast metal. His clothing? The general came with "real leather hat, real buckskin jacket and pants, and even REAL leather boots. Not fake leather, REAL Leather. Would Custer be caught in fake leather?"

Recall PETA's inspired fury at the pain it sees embodied in every piece of leather? When the FTC sought comments on its labeling guidelines, PETA asked that consumers of leatherwear be warned, "Animals Suffered to Make This Product." Its mission can seem radical, thankless, or both; cost no object, consumers won't normally prefer vinyl siding to wood, Formica to stone or, as PETA would, "pleather" to leather. But seen another way, its position is familiar and conventional: To PETA, whether shoes, belts, or briefcases are made of leather or something else *matters.*

"I sat by a fountain and read for a while," says a character in science fiction writer Joe Haldeman's novel *Forever Peace.* He holds the book, experiencing "the heavy yellowed paper, the feel and musty smell of the leather. The skin of an animal dead more than a century, if it was real leather." *If it was real.* Why wonder? Why not just take it for whatever it is? But to Haldeman's character it matters, and it's not so strange that it should. We live in a world of things and, deep down, we want to know where they come from, what they *are.* Where a manuscript or painting originates, its "provenance," helps fix its value to auction galleries and art museums. Is it silver or silver plate? Is it a *real* Picasso? Does it come from cow or petrochemical plant? *We want to know.* When in 1910 the *Wall*

Street Journal recorded Du Pont's purchase of Fabrikoid, it noted that the company's "'near leather' is surprisingly like the genuine article." Somehow, within the material itself, the question almost irresistibly beckoned.

Computers reduce blood, wax, gold, weather maps, and poker hands to the same zeros and ones of digital imagery. Is it time, finally, to get our hands on things again that are neither plastic nor virtual? At a textile arts exhibition in Korea a few years ago, sculptor Warren Seelig, of Philadelphia's University of the Arts, thought out loud about what he called "material meaning." Young artists, he said,

> have been distanced from the physical world, slowly at first, but in recent decades with increased rapidity. The materials we experience every day, in our homes and workplaces and in the objects we live with, are more often than not synthesized or reconstituted—with their surfaces neoprene-coated, plasticized, veneered, laminated, plated, etc.—and are thus so homogenized as to be ultimately unrecognizable. Our bodies are hermetically sealed off from the outside world in climate-controlled homes, schools, office buildings and automobiles. We peer out at the world through triple-glazed windows, and more and more we experience physical reality through the lens and filter of a high-resolution plasma television screen. Recognizing that we are all enveloped in this atmosphere, it is not surprising that artists more than ever are choosing materials like body fluids, animal carcasses, hair, bone, dirt and other kinds of organic and industrial debris, not only to shock the viewer, but also to resensitize us to the carnal and to the world of the physical. Exposure to the primal reality of raw material, whether it is oak, granite, iron, gold, linen or silk, can be a startling revelation to the many whose closest contact with the natural world comes through wood-grained Formica.

The primal reality of raw material? "We are instinctually drawn" to such materials, Seelig says, "out of a profound physical and psychological need."

Some years ago appeared a book, *Wooden Boat Renovation,* that explained wood's challenge to the prevailing fiberglass orthodoxy. Its author, Jim Trefethen, had once worked for a utility company in a big city office building. That was back in the mid-1960s, "so I wore the obligatory textured polyester suit and knitted polyester shirt. My socks were made of nylon and my shoes of Corfam. . . . In my fluorescent-lighted office with its asphalt-tile floor, I worked at a metal desk that had a plastic-laminate top, and I sat in a steel chair upholstered with Naugahyde." Now, though, he reported from his new life in the 1990s, he wore cotton and wool, and his shoes were of top-grain cowhide. There'd been "something terribly wrong with life in an artificial environment." Something indefinable, maybe, "but when it's not there we miss it and long to get it back." Trefethen was no philosopher; he was writing about boats. But he was prepared to assert as fact that people preferred natural materials.

Proof? No, but he did offer one small suggestive dollop of evidence. "In today's corporate environment," he wrote,

> wood is used as a reward for achievement and as a sign of status. . . . When [executives] finally reach the very top of their profession, they will, likely as not, be ensconced in a wood-paneled office with a wooden desk and a wooden door. On the oak floor will be a wool carpet, and their mahogany chair will be upholstered with real leather. They have arrived, and in doing so, they have earned the right to have real things.

Real things, the ultimate reward. But intrinsically so? Or do wood and leather seem desirable only at the whim of fashion? University of Iowa textiles expert Sara Kadolph has observed that for a time vinyl was all the rage—until the reaction set in with *The Graduate,* the death of Corfam, and the fallout in values from the tumultuous 1960s. Confronted by synthetics, one trade journal writer noted, the tanners created "the image of 'leather, the superlative material nature grows for you.'" Was it, then, all just good PR? Or was "something already in the air—people suddenly wanting to live among natural products and not synthetics?" For him, at least, a reaction had set in to lives grown "so unnatural in every way" that "antagonism . . . against materials grown in the test-tube" stood ready to be exploited.

Trefethen's real-things-as-reward principle may be no immutable law of nature. But it's likewise true that, whatever new materials issue from the lab and whatever the sway of fashion, the demand for wood, wool, gold, and leather never abates for long. Late in *À Rebours,* that delirious celebration of the artificial, Des Esseintes resolves to have his books reprinted on hand presses, on special handmade papers, and painstakingly rebound "in irreproachable bindings of old silk, of embossed ox-hide, or Cape goatskin." There was no dearth of faux leathers he could have picked instead; but for Des Esseintes, Prince of Faux, only those exquisite natural materials would do.

Ernest Dichter, the motivational research consultant, described what he called a "dream of naturalness," a nameless, primeval *something* "from deep down, from the 'animal within us.'" Psychologists speak of that disease of modernity—alienation. Well, "getting back to the real things, the things we knew about in our animal past, gives us a feeling of security and is a cure for this alienation." "Real" things had depth, they were imperfect, they changed—they lived. Dichter found that while Corfam and Formica—he mentioned both by name—had advantages over their natural counterparts, "consumers keep going back to nature as the ideal. We speak about 'genuine leather,' 'real wood,' and 'real soap.' In many of our studies we find again and again that when people are questioned about their favorite products, the natural ones come first to mind."

Dichter wrote that in 1964. But four decades and several swings of the fashion pendulum later, the Western Red Cedar Lumber Association was trading on kindred sensibilities in an ad campaign that poked fun at a fictional rival it called Plastibord:

> From 50 feet away, it almost looks natural! Your fences, sheds, gazebos, arbors, planters, pergolas, pagodas—yes, all the comforts of gnome will glitter with plastic perfection when they're lovingly crafted with PLASTIBORD! But wait, there's more . . . your standout PLASTIBORD creations are guaranteed never to blend into the surrounding landscape. Ever!

The Red Cedar people gave Plastibord its own slogan: "With stuff like this, who needs Nature?"

Morris Kaufman has written of a "subjective resistance . . . very deeply based" to materials that would replace wood, wool, leather, and vegetable dye, chalking it up to innate antipathy to the new. But maybe some of it inheres not to the newness of such materials but to their distance from nature? And perhaps, as well, despite the best efforts of so many scientists, artisans, and engineers, to their inner lifelessness?

Products of mass production, wrote David Pye, "lack depth, subtlety, overtones, variegation" and what he famously called "diversity," meaning the richly varied effects discernible at various scales and distances. Fine craftsmanship, meanwhile, "imports into our man-made environment something which is akin to the natural environment we have abandoned." Leather, of course, needn't be "imported" from nature; it originates there. And at least when not too gussied up by pigment and alien surface grain, it shows itself off best up close, in just those intimate precincts of eye and fingertip where the appeal of synthetics vanishes.

In *What Is Life?*, his classic meditation on the boundaries between physics and biology, written a decade before Watson and Crick deciphered the structure of DNA, quantum physicist Erwin Schrödinger theorized what the genetic material might be like. He imagined an "aperiodic crystal" whose very irregularity would leave it able to carry genetic information; nothing monotonously regular could possibly work. "The difference in structure," he wrote,

> is of the same kind as that between an ordinary wallpaper in which the same pattern is repeated again and again in regular periodicity and a masterpiece of embroidery, say a Raphael tapestry, which shows no dull repetition, but an elaborate, coherent, meaningful design traced by the great master.

Midcentury German critic Walter Benjamin wrote of the "aura" of an original work of art, that singular quality of the living here-and-now residing within it. Can leather be seen as a work of original art? Does it lose Benjamin's "aura," suffer Schrödinger's "dull repetition," when simulated in 10,000 yards of Lorica?

Replicas, it's been said, do "not diminish the esteem in which the work from the hand of the master is held." Perhaps the abiding hold of materials like leather is that the "master" is Nature itself.

11
Vera Pelle

He comes from five generations of leather people. When he was 16 he worked in the basement of his dad's Leather District shop in Boston, sorting skins. His name is Howard Shrut, and his company, Shrut & Asch, he calls "the only domestic supplier of kidskin in the US" still left. Once, it employed 30 people. "I used to love to go into a shoe store and say, 'That's my color, that's my customer.'" He still sells half a million square feet a year, versus 16 million in the company's best years, in the 1970s. But now they're down to two full-time employees, including himself, and one part-time. Their offices are in the shadow of Fenway Park, expressway traffic whizzing by almost within earshot. In the back room a gridwork of wooden shelves hold rolled-up skins. This season a lot of them are bright mauves. He pulls one down. Ask for one the barest bit thinner and he can pick it out without a gauge, just from the feel of it between his fingers. He speaks of a skin's "roundness," how it feels as you fold it into a tight curve.

One day back in the 1970s he went to take an order from one of his biggest, steadiest customers. In his order book he wrote down 10,000 feet of navy blue, then looked up. "And black?" Black was a given; however large or small a customer's order, black would be part of it. But this time his old customer hemmed and hawed, finally spit it out: "Well, we're importing a little of it now."

That, in Shrut's mind, was the beginning—the decline of the American leather industry. A few years earlier, imports from Spain and Italy were already contributing to the drop in leather prices that helped undo Corfam. The trend accelerated. Today the American leather industry is moribund. A recent annual meeting of Leather Industries of America in downtown Boston was a grim affair, attendance maybe 20, down from 75 in half a generation, its president and most everyone else all singing the same lament. The United States still produces millions of raw hides, but they go overseas, where they're tanned and finished, made into shoes and wallets that are shipped back here. Of shoes bought by Americans it's not 90 percent that are imports, says LIA president Charlie Myers, not even 95 percent, but about 98 percent. Shoe manufacturers in India, Indonesia, Mexico, Vietnam, and China draw tanners. The tanners draw chemical houses and tanning machinery makers. In the end they've got all they need in their own backyard; there's little reason to buy American. The same goes for Europe; the French tanners, the Spanish tanners, are ghosts of what they once were. These days, says Shrut, head over to China if you want to see what venerable old American leather towns like Gloversville, New York, or Peabody, Massachusetts, were like in their prime.

The way the whistles blew—that's what one worker remembered best about Peabody. During World War II 20,000 people worked in its tanneries, more than 100 of them making leather bound for nearby shoe-making towns like Brockton and Lynn. It was tanneries on every corner, canals polluted, the work wet, dirty, and hard—"foul odors and multicolored crud," as the narrator says in a documentary, *Leather Soul*, about the old days of Peabody leathermaking. But the odors and crud spelled prosperity and steady jobs. "Like music to my ears, like Mozart at his best, all the factories blowing their whistles," recalled the worker.

The film, produced by Joe Cultrera and narrated by the inimitable radio personality Studs Terkel, appeared in 1991, by which time the tanneries were all but gone, jobs down to a few hundred, old tannery buildings turned into condos. It opens with scenes of cracked window glass and peeling paint. We see a close-up of a freckle-faced boy with full red lips, maybe 11 years old. "Ever hear of the leather industry?" an off-screen interrogator asks.

"Un-unh."

"Ever hear of tanning?

"Nope." His face scrunches up in embarrassment, or incomprehension.

What happened to the American leather industry? Tannery workers and owners alike offer their opinions. One says it's the government to blame, because it failed to impose import duties.

For another it's pollution, and the industry's costly, too-late efforts to remedy it.

A third blames the unions.

"Plastics," says a fourth. It was "those imitations, that's what put the leather business out."

But plastics were only incidental to American leather's decline. In a look back at the industry's last half-century presented at the 100th anniversary meeting of the American Leather Chemists Association in St. Louis in 2004, John Koppany told how, with sole leather's demise, it seemed certain that shoe, upholstery, and garment leather would be next. It didn't happen:

> Remember the Corfam scare? E. I. Du Pont's shares soared as the best U.S. men's shoe manufacturers were beginning to sell their top-line fashion dress shoes made in Corfam . . . and then fortunately their shares plummeted as well when it was proven that Corfam shoes suffered from exaggeratedly good memory and one's feet had to go through molding the 'new' shoe on a daily basis.
>
> Remember Alcantara, the "magic" Japanese material that was going to be the death of suede goat, calf, sheep and splits. . . . Well, at the moment the world is seeing record priced splits in demand for garments and upholstery.

All told, "some battles were lost and others won."

In fact, it's misleading to speak of a single "leather industry"; when leather clashes with synthetics, it does so on distinct battlefields, in commercial niches only indifferently linked. Like shoe soles, which it lost almost entirely; 1 percent of shoes reaching America have leather soles. And shoe uppers, where it defeated Corfam and continues to prosper. Garment leather? Car interiors? Leather goods, like wallets and briefcases? Each a battle among materials. Koppany showed a chart representing growth or decline in this niche or that, some arrows up, some down. On balance, though, leather has held its own. Even if American and European leather have not.

It's the explosive growth of Asian economies that earns the front pages, of course, not our quieter story of dueling materials. Scan industry technical journals and many contributors bear names like Ramkumar or Ramalingam, work at the Central Leather Research Institute of Chennai, formerly Madras, India. China has 20,000 shoe factories, whole districts given over to tanning. So if we think globally for the moment and don't let our gaze settle too long and too painfully on abandoned American tanneries and shoe factories, it's plain that leather itself, however challenged by synthetics, endures.

But will it always?

In an unpublished study of Fabrikoid, historian John Kenly Smith observed that Du Pont's decision to enter the artificial leather business "was not the opening move in a grand strategy to make Du Pont the pioneer in synthetics," though it had that effect; artificial leather was an outlet for the company's excess nitrocellulose capacity, nothing "grand" about it. Similarly, we're wise not to impute over-grand motives to imitators of leather through the years. Few, we may suppose, were moved by a consuming personal quest to imitate nature; or by some *Frankenstein*-like determination to erase the bounds between the living and nonliving worlds; or, more improbably yet, to usurp the work of God. But that their work can be seen in such flat prosaic light doesn't mean that the technicians, artisans, and engineers who did it were not sometimes brought up short in wonder by what they attempted; Corfam and Ultrasuede were not the products of withered imaginations. For 150 years and more, inventors have bruited claims to the world that have always, in some measure, fallen short. But they've never stopped trying:

1856: "Imitations of Leather," British Patent 1862

1904: "Production of Substances Resembling Leather," U.S. Patent 750,371

1919: "Leather Substitute," U.S. Patent 1,305,621

1955: "Leatherlike Products and Preparation of Same," U.S. Patent 2,715,588

1966: "Artificial Leather," Japanese Patent SHOU57-59353

1992: "Process for Converting Composite Imitation Leather Into Sheet Material Similar in Appearance to Natural Leather," U.S. Patent 5,290,593

2005: "Suede-like Fabric," U.S. Patent 6,878,407

It is this dogged quest that, coupled to the enduring pleasures of leather itself, runs down the spine of this book. And is there not something noble in it, this courage to make one more try, with something new, maybe better? One harsh way to see the maker of imitation leather is as a purveyor of inferior fakes. But another is as social benefactor—bestower of materials more abundant, cheaper, sometimes even more ecologically responsible. That's no mean legacy.

Eulogies for leather are premature. But might it one day join ivory, ebony, and gutta percha among natural materials vanished from daily life? Certainly it's being pressed in new ways. Aesthetically, Naugahyde and Corfam offered little threat. Ultrasuede and Amaretta surely do; luxurious look and feel contribute to their appeal as much as breathability or tear strength.

Kansei, not just *Risei.*

Miyoshi Okamoto made that distinction at a conference in Otsu, Japan, in 1991, two decades after his Ultrasuede work, as he considered the many fine *shingosen* textiles evolving from it. *Risei,* he explained to his international audience, referred to ordinary functional measures like tensile strength. *Kansei* meant "sensibility"; but the word was so steeped in a Japanese cultural context that Okamoto offered complementary definitions for it, 16 of them. Together, they suggested materials valued no longer only for utilitarian qualities but for aesthetic, sensual, and emotional ones. In *The Substance of Style,* Virginia Postrel likewise emphasizes the aesthetic dimensions of materials: "Our sensory side is as valid a part of our nature as the capacity to speak or reason, and it is essential to both. Artifacts do not need some other justification for pleasing our visual, tactile, emotional natures." With its warmth, sheen, and hand likely targets for future imitators, leather thus stands threatened in yet new ways.

But if leather is to succumb to ever-better lookalikes and feelalikes and become as alien to our lives as ivory, might we try to recall, before it's wholly lost to us, what it meant to so many for so long?

They gather over lunch, on the second floor of an old villa in Ponte a Egola, Italy, working on behalf of leather. They represent the 31 tanneries of the Consorzio Vera Pelle Italiana Conciata al Vegetale, a consortium that promotes Italian vegetable-tanned leather—leather with roots early in human society, made with tree bark and plants, the way people did it a thousand years before Augustus Schultz ever dreamt of chromium salts.

We are in one of the three beating hearts of the Italian leather industry. Down south, near Naples, is Solofra. Up north near Venice is Arzignano, its leathermaking history going back to the 15th century, its big tanneries turning out leather for upholstery and car interiors. Here, in between, on the dusty Tuscan plain between Pisa and Florence, is Santa Croce sull'Arno, home to perhaps 500 mostly smaller tanneries. This is Tuscany, yes, but not the Tuscany of the tourists, who are more apt to be found admiring Brunelleschi's Dome in Florence or slogging their way across Pisa to the Leaning Tower; few stop at the tiny San Miniato station where, after taking the Florence-bound local from Pisa, past ragged residential districts and irrigation canals, stopping at little towns every few minutes, you get off to visit leather country.

On a map of Italy the Santa Croce area covers no more than a dot—100 square miles, a tenth the size of Rhode Island. But within this area, more than 15,000 people work in leather. Tanneries on every cul de sac, signs for *conceria*, tannery, and *pelle*, leather, on every corner. No sprawling complexes here; these are mostly small operations, maybe a dozen employees each, none more than a minute or two's drive from its neighbors; the midday traffic—a double rush hour in siesta-hungry Italy—is crowded with tannery workers.

In Santa Croce itself, leathermaking mostly means chrome. But across the Arno and down the road a few miles is Ponte a Egola, the hub within a hub of Italian veg-tanning. Don't look for pits, year-long tannages, or men leaning over their beams here; this is not the middle ages. The great wooden drums revolve just as they do for the chrome tanners across the road or outside town. Conveyors silently lift finished hides through second-floor drying rooms. But as we saw many chapters earlier, and as the Consorzio takes pains to highlight, veg-tanning makes for a not-so-subtly different leather.

Formed in 1994, the consortium is financed in part by vegetable extract producers. It's these extracts you see piled up beside the Tempesti tannery, on via del Cuoio, Leather Street. Twenty-five-kilogram plastic bags of fine powder the consistency of wheat flour or cocoa mix—tara from Peru, mimosa, rust-red quebracho. Veg-tanning takes longer—a month or more versus four or five days for chrome tanning. And it costs half as much. But the result, says the Consorzio, is leather as close to its roots in nature as a man-made product can be. "Leather tanned using natural vegetable tannins extracted from trees," it declares, "becomes softer, more ductile and durable. Colors become deeper through usage and the passage of time, as if tanned by the sun. A true marvel of nature."

Listen to Stefania Miniati, the consortium's international promotions coordinator and she'll tell you that chrome leather is not so good environmentally, that veg-tanned leathers more readily return to the earth. "I didn't know anything about leather before; I came from Florence," she says, as if it were a million miles away instead of a morning's commute by car. But at the university, where she was studying economics, her professor knew all about Tuscan veg-tanning. It became her thesis topic and she among its biggest fans. "You have to touch and smell it," she says. "The smell reminds you of trees, of wine." Her colleague, Valentina Sgherri, tells how belts and handbags of veg-tanned leather change with time, reflecting the life experience of the user; otherwise identical articles look different after a few years. But always deep, dusky—and natural.

Here in Tuscany, where *decalcinazione* is bating and *conciatura* is tanning, the tanneries are never very large. Each has its few great wooden drums, turning five or six times a minute, barely contained by the spartan industrial structures that house them, practically scraping the ceiling. Most companies specialize. One splits leather. Another, not much larger than a big American garage, makes sole leather; half of Europe's leather soles, all veg-tanned, come from the area. A third—its workshop up from the street, closed off from prying eyes—specializes in *tamponatura a mano*, the hand daubing of leather in intricate patterns. The Tempesti tannery, a little bigger than some, works with the part of the hide known as the shoulder. Puccini Attilio, presided over by Stefano Casella, an obliging man with long scraggly hair, deep-set eyes, and a dark, wispy beard, makes leather with wild splashes of color; one resembles a shower curtain replete with water droplets. If Italian leather is dead or dying, as some might say who have seen its better days, you wouldn't right off know it from Ponte a Egola.

Rows of burnished brown leather seats, banks of them in threes and fours, sitting on frames of functional black steel tubing. These aren't light, supple skins pulled tight across foam cushions, but broad sweeps of thick vegetable-tanned leather, lightly padded underneath. Their deep rounded furrows make you feel like you're sitting on a saddle, not a chair.

And they're not in some executive board room, either, but at the airport, for anyone to plop down into. Stamped on the back, AEROPORTO DI ROMA, they show the stigmata of daily use, scratches and gouges from who knows how many errant keys, zippers, or jeans rivets, splotches and discolorations. And yet, polished by use, they remain rich and sumptuous. They pick up the ceiling lights, gleam from every contour. They are stunningly beautiful. This is Italy. Beauty is the national industry.

In Italy, compared to the rest of Europe and North America, the hemorrhaging of centuries-old leather industries at the hands of China and India can seem remote. Next to that of France or England, say, and certainly of America, Italian tanning is vibrant, $5 billion worth a year. While British shoe manufacturers annually produce just 30 million pairs, American 60 million, and French about 75 million, Italy makes 335 million—fully half that of the United States in its post-World War II heyday. Italy remains a style center; Gucci, Versace, and Ferragamo still sparkle in the fashion firmament. Italian tanning machinery makers supply the world; many machines going into Chinese tanneries are made in Italy or copied from Italian designs. "The Italian leather industry," a trade journal noted in October 2002, "has been likened to a hornet; by all aerodynamic laws, it cannot possibly fly, but as it does not know this, fly it does."

But its flights have become more erratic; you can hear the engine sputtering from the ground, Italian leather losing altitude. Sales of both leather and footwear are down and, even in good years, make up a smaller fraction of the world's production; in 2002, Italy made one-twentieth as many shoes as China. "Deep-Seated Concerns for the Future," ran the headline for a review of Italian tanning three years later. The Solofra district had long struggled. Arzignano was down. Santa Croce? It, too. From 2002 to 2003, sales by Consorzio tanneries fell from 156 million euros to 130 million, about 17 percent, though they've since bounced back.

The Consorzio has made little progress cracking the U.S. market, laments Stefania Miniati; in America the talk always comes down to price—sometimes, it can seem, nothing *but* price. For four years they solicited business in Japan and made headway. "There they believe in tradition. They love our product. Oh, for days and days you have to explain things.

But if you can explain the special characteristics of the product, they go for it," almost regardless of price. In America, on the other hand, the Italian tanners faced the tyranny of the numbers.

China and other developing regions *have* the numbers—low labor costs—and are trying to break into the veg-tanning market. But Tuscan leather, Miniati as much as says, can be made only in Tuscany. "You need particular artisans," artisans who know the drying winds, and these, she says, they'll never have in China, India, or Mexico. "They don't have Tuscany. They don't have the weather. They don't have the wind."

But they're trying, says Paolo Testi, vice-president of Conceria La Bretagna, one of the Consorzio's 31. The Chinese, he reports, come to Lineapelle and other leather fairs, lure Italian technicians, Italian tanning machinery companies, all to help them make leather the Italian way. But he can still tell the difference, says Testi, between Tuscan veg-tanned and Chinese, from "the depth of finish, the hand, the softness." That's the party line, of course; in fact, his confidence is neither absolute nor blind. Threatened? Of course, they are. "That's one reason we are here." That's what the consortium has come together to protect.

Testi is among the half dozen of them ranged over toward one end of the great wood table that fills the airy second-story room in the Ponte a Egola mansion. Beside turquoise shutters swung open from the bottom, vertical blinds bear a familiar animal skin logo. A mural on one wall shows tannery hands holding hides aloft like banners. They are not the Beautiful People, these men and women trying to preserve *vera pelle Italiana conciata al vegetale.* They're a little ragtag, smoking their cigarettes and cigars, chomping down their pizza, batting out ideas for an upcoming exposition. One wears a sweatshirt that, unaccountably, bears the legend BONFIRES. Another, in motorcycle gear and trim white goatee, looks like he just got off his Harley. For a couple of hours they go at it. One of them, a photographer with dark hair held in a ponytail and fire in his eyes, at one point steps to a large easel-mounted pad of paper and sketches, in blue marker, Tuscan hands making Tuscan leather.

"When you touch this leather, you feel good," says Testi. "You don't know why, but you do." A competitor, Puccini Attilio, issues a plain folder with samples of dense, hard leather—much like what went into the seating at the Aeroporto di Roma—for suitcases, bags, and belts. The colors—

cioccolato, arancio, assuro—reach a mile down beneath the surface. No Naugahyde, no Amaretta, ever looked like this.

> Our products are made of "Vera Pelle" Real Bovine Leather. Not always it is possible to eliminate all the small imperfections and veinings, not even after a very accurate skin preselection. So these should not be considered as defects but peculiarities of a very Natural Product.

Call it hype. Call it love.

"Robust marketing" is how a trade journal described the Consorzio's methods. Its brochures are minor masterpieces, its video presentations all color and hypnotic rhythms. A couple of times a year the Consorzio goes to Tokyo, the Guggenheim Museum in New York, Centre Pompidou in Paris, or Milan to put on seasonal shows for journalists, designers, and leather goods makers—holding the aesthetic high ground, trading off nature's cachet.

"Natural Sensations," they call these fashion trend reviews. "In a journey of discovery that winds through Natura and Cultura," one brochure rhapsodizes in Italo-English, "the research of Natural Sensations explores the essence of moments where the moving force of Nature is transformed into the myriad expressions of Culture." Then it's close-ups of flowers, seaweed and clamshells beside the beach, seed pods and haystacks, water trickling over rocks, explosions of color and form. Natural forces, natural textures. No human taint here; no gorgeous Italian models to distract from the impression. Just nature at its most alluring that this leather—so close to nature, *in* nature, *part* of nature—is supposed to represent.

Must we fault these aesthetic assaults of image and word? Certainly they pander to easy nostalgia—for tradition, Tuscany, artisanal prowess. But seen another way, they constitute a reply to those dreary, utilitarian claims most ordinary materials make on the world. Through our senses, beautiful materials—all materials, really—reach into our lives: We touch them, they touch us. For these Tuscan tanneries, these few hundred men and women, this one ancient, natural material bestows beauty, grants pleasure, cries for lasting life.

Sources and Acknowledgments

While the research for *Faux Real* rests in part on a familiar bibliographical edifice of published books and articles, it draws at least as heavily on interviews, archival records, private papers, and tours of industrial facilities. It relies on journal articles, technical glossaries, U.S. government reports, and scholarly dissertations; on ads, annual reports, brochures, popular articles, patents, websites, and online forums; on fiction that bears on its broader themes; and on my own experience as an amateur leather craftsman. In the pages that follow, I try to leave readers with a sense of how *Faux Real* came to be; through this reverse engineering, I suppose, a reader could come away with quite another view of the whole subject.

In the text, I've written about a few patents in detail but the number of patents devoted to natural and imitation leather runs to many thousands. So I have not tried to methodically survey them or to uncover patterns among them—as, for example, Amanda Lindsay has done in her doctoral dissertation devoted to microfibers, cited below. Thanks to Theresa Riordan, for her help in navigating the patent system.

While writing this book I reviewed substantial runs of the following specialized and trade publications: *American Craft*; *American Shoemaking*; *Boot and Shoe Recorder*; *Du Pont Magazine*; *Japan Textile News*; *Leather Crafters and Saddlers Journal*; *Leather Manufacturer*; *Journal of the Ameri-*

can Leather Chemists Association; *Journal of Coated Fabrics*; *Journal of the Society of Leather Technologists and Chemists*; and *World Leather*. From these I typically cite only the most significant articles or those upon which I most heavily relied.

Mentioned in this book are products and materials that may or may not be around five years from now, or 50. In the notes that follow, therefore, I don't normally cite routine press releases, annual reports, commercial websites, informational brochures, and advertising of the sort that crowds around any commercial product. It should be clear, though, that these helped me understand how these products were viewed by their manufacturers and how they were presented to the world.

Above all, I am indebted to many men and women, cited by name below, who spoke to me about leather and its imitators, in person, by phone, or both; who let me visit their tanneries, shops, and other places of business; engaged in extensive email correspondence; or otherwise shared with me their knowledge, opinions, and recollections. None of them, of course, are responsible for any errors, omissions, or misinterpretations I may have made.

SOURCES AND ACKNOWLEDGMENTS is structured as follows:

Leather

- History, Culture, Lore
- Leathermaking
- Tanning
- Leather Goods, Products, Craft
- Global Markets

Faux Leather

- Pyroxilin
- Vinyl
- Corfam
    - Unpublished Sources
    - William Rossi Collection
    - Published Overviews
    - On the Eve of Corfam
    - Technical Articles
    - Other Publications

LEATHER

My knowledge of leather, its manufacture, its working into products, and its history have benefited greatly from visits to industrial and commercial facilities where I gained much thanks to the knowledge, patience, and goodwill of the following people: Mike Whalen, Griffin Industries, Cincinnati; Richard C. Larochelle, Irving Tanning Company, Hartland, Maine; Nick Cory, Kadir Donmez, and the late Randy Rowles, Leather Research Laboratory, University of Cincinnati; Stefania Miniati, Valentina Sgherri, and Paolo Quagli of Tuscany's Consorzio Vera Pelle Italiana Conciata al Vegetale; Giorgio Tempesti of Conceria Tempesti; Stefano Casella, Puccini Attilio tannery; and Howard F. Shrut, Shrut & Asch Leather Company, Boston.

Members of the American Leather Chemists Association, whose ranks I joined while working on this book and whose 100th anniversary conference in St. Louis I attended, were an unfailing source of contacts and information. Special thanks to Waldo Kallenberger, kingpin of the ALCA leather discussion forum, whose fascination with all things leathery has been contagious, and sometimes provocative, across the three years during which I stepped into his world and that of his ALCA colleagues. Many thanks, as well, to Jean Tancous, who let me spend the day with her in her basement laboratory, happily talking leather all the while; and to Mike

Redwood for his thoughtful insights about the industry into which he was born and to which he has contributed so much.

I have also profited from interviews with Jim Bates, Sergio Castro, Gene Killik, John Koppany, David Rabinovich, and Paolo Testi. Thanks, too, to Leather Industries of America members Charlie Myers, Jack Mitchell, Lisa Howlett, and Jean Ann Firestone; to Steve Lubar and Peter Liebhold for their welcome to the splendid facilities of the Smithsonian and for showing me the special boots made for the 1876 exhibition and the Mahrenholz correspondence going with them; and to Walter Lind for right there, in the middle of Lake Winnipesaukee, New Hampshire, remembering *Leather Soul*, the excellent documentary about Peabody, Massachusetts. Finally, thanks to Eleanor Brown, research chemist with the Eastern Regional Research Center, Wyndmoor, Pennsylvania, for reviewing several sections of this book; any errors that may remain are, of course, my doing, not hers.

In its millennia at the right hand of men and women, leather has generated a vast literature. Herewith is a sample:

History, Culture, Lore

American Leathers. New York: American Leather Producers, 1929.

Cameron, Esther, ed. *Leather and Fur*. London: Archetype Publication Ltd., 1998, pp. 1-56.

Clarkson, L. A. "The Organization of the English Leather Industry in the Late Sixteenth and Seventeenth Centuries," *The Economic History Review*, New Series, vol. 13, no. 2, 1960, pp. 245-256.

Cochrane, Charles H. *Modern Industrial Progress*. Philadelphia: J. B. Lippincott, 1911, pp. 527-537.

Dana, Richard Henry. *Two Years Before the Mast*. New York: Signet Classics, 1964.

Davis, Charles Thomas. *The Manufacture of Leather*. Philadelphia: Henry Carey, Baird & Co., 1885.

Diderot, Denis. *A Diderot Pictorial Encyclopedia of Trades and Industry*, vol. 2, Charles Couston Gillispie, ed. New York: Dover Publications, 1993.

Donham, Richard. "Problems of the Tanning Industry," *Harvard Business Review*, July 1930, pp. 474-481.

Ellsworth, Lucius F. *The American Leather Industry*. Chicago: Rand McNally, 1969.

Frankfort, Lew. *Portrait of a Leathergoods Factory*. New York: Coach Leatherware, 1991.

Leather Facts. Peabody, MA: New England Tanners Club, 1994.

Leather Soul: Working for a Life in a Factory Town. Documentary. Directed by Joe Cultrera. Narrated by Studs Terkel. Picture Business Productions, 1991, 46 minutes.

Les Tanneurs de Marrakech. Documentary. Directed by J. Aityoussef. Service du Film de Recherche Scientifique, 1967, 21 minutes.

Lock, Charles G. Warnford, ed. *Spons' Encyclopedia of the Industrial Arts, Manufactures, and Raw Commercial Products.* London: E. & F. N. Spon, vol. II, 1882, pp. 1213-1228.
"Nothing Takes the Place of Leather: A Brief History of Leather and a Description of Tanning." Booklet. New York: American Sole & Belting Leather Tanners, 1924.
Pritchett, V. S. *Nothing Like Leather.* New York: Macmillan, 1935.
Quilleriet, Anne-Laure. *The Leather Book.* New York: Assouline Publishing, 2004.
Redwood, Mike. "Nature's High Performance, Breathable Material," *Leather Industry Companion,* from a lecture given to the "Survival—90" conference at the University of Leeds, March 1990.
——. "Technical Leathers in an Active World." Transcript of lecture given in Shanghai, October 6, 1993.
Spiers, C. H. "Sir Humphrey Davy and the Leather Industry," *Annals of Science,* vol. 24, no. 2, June 1968, pp. 99-113.
Spindler, Konrad. *The Man in the Ice.* New York: Harmony Books, 1994.
Thomas, S., et al. "Leather Manufacture Through the Ages," proceedings of the 27th East Midlands Industrial Archaeology Conference, October 1983, 35 pp.
Thomson, Roy. "Leather Manufacture in the Post-Medieval Period with Special Reference to Northamptonshire," *Post-Medieval Archaelogy,* vol. 15, 1981, pp. 161-175.
Tree, Christina, James Sutton, and John Moynihan. "Leather Through the Ages," *The Leather Manufacturer,* June 1973, 14 pp.
Wagner, Rudolf. *Wagner's Chemical Technology 1872.* New York: Lindsay Publications, 1988, pp. 508-523.
Waterer, John W. *Leather in Life, Art and Industry.* London: Faber and Faber Limited, 1943.
——. *Leather and the Warrior.* Northampton: Museum of Leathercraft, 1981.

Leathermaking

Bailey, David G., et al. "Leather," *Kirk-Othmer Encyclopedia of Chemical Technology,* 3rd ed., vol. 14. New York: John Wiley & Sons, 1981.
Fuchs, Karlheinz H. F. *Chemistry and Technology of Novelty Leathers.* Rome: Food and Agriculture Organization of the United Nations, 1976.
Heidemann, Eckhardt. "Leather," *Ullmann's Encyclopedia of Industrial Chemistry,* 5th ed., vol. A15. New York: Wiley-VCH, 1990.
"Mini-Symposium on Soft Leathers," 71st Annual Meeting of American Leather Chemists Association, June 23, 1975.
O'Flaherty, Fred, William T. Roddy, and Robert M. Lollar. *Chemistry and Technology of Leather,* 4 vols. New York: Reinhold, 1956-1965.
"Overview of Leather and Parchment Manufacture." Online at: *Koninklijke Bibliotheek, National Library of the Netherlands. http://www.kb.nl/cons/leather/chapter1-en.html.*
Portavella i Casanova, Manuel. *Leather . . . this natural wonder. La Piel . . . este prodigio naural.* Vic (Barcelona): Colomer Munmany, S.A., 2000.
Reed, R. *Ancient Skins, Parchments, and Leathers.* London: Seminar Press, 1972.
Roddy, William T. "The Wondrous Inside World of Leather," *Boot and Shoe Recorder,* September 1, 1966, 8 pp.
The Story of Leather: A Trip Through a Modern Leather Plant. Girard, Ohio: Ohio Leather Co., ca. 1949.

Taeger, Tilman. "Progress in Leather Chemistry: What Kind of Milestones Are to Be Expected?" *Journal of the American Leather Chemists Association*, vol. 91, 1996, pp. 211-225.

Tancous, Jean J. *Skin, Hide and Leather Defects*, 2nd ed. Cincinnati: Leather Industries of America, 1992.

Wood, J. T. "Fermentation in the Leather Industry," *Journal of the Society of Chemical Industry*, March 31, 1894.

Wood, Joseph Turney. *The Puering, Bating & Drenching of Skins*. London: E. & F. N. Spon, 1912.

Tanning

Atkinson, J. H. "Vegetable Tannage—Past, Present and Future," *Journal of the Society of Leather Technologists and Chemists*, vol. 77, pp. 171-173.

Bienkiewicz, K. J. "Leather—Water: A System?" *Journal of the American Leather Chemists Association*, vol. 85, 1990, pp. 303-325.

Brodsky, Barbara, et al. "Collagens and Gelatins," in *Polysaccharides and Polyamides in the Food Industry*, Alexander Steinbüchel and Sang Ki Rhee, eds. Weinheim: Wiley-VCH, 2005, pp. 119-128.

Covington, A. D. "Theory and Mechanism of Tanning: Present Thinking and Future Implications for Industry," *Journal of the Society of Leather Technologists and Chemists*, vol. 85, 2001, pp. 24-33.

Daniel, Richard. "Back to Basics," *Leather Basics*, March-April 2003, pp. 49-50.

Dewhurst, J. "Oil Tan Buff Leather—Man's First Leather?" *Journal of the Society of Leather Technologists and Chemists*, vol. 88, pp. 260-262.

Harlan, J. W., and S. H. Feairheller. "Chemistry of the Crosslinking of Collagen During Tanning," *Advances in Experimental Medicine and Biology*. New York: Plenum Press, 1977, pp. 425-440.

Haslam, E. *Chemistry of Vegetable Tannins*. London: Academic Press, 1966.

Hatchett, Charles. "On an Artificial Substance Which Possesses the Principal Characteristic Properties of Tannin," *Philosophical Transactions of the Royal Society of London*. London: The Royal Society, 1805, pp. 211-224.

Moore, W. R. "The Structure and Properties of Natural and Synthetic High Polymers," *Journal of the Society of Leather Trades' Chemists*, vol. 50, no. 3, March 1966, pp. 94-109.

Procter, H. R. "Chrome and Iron Tannages," in *Principles of Leather Manufacture*, 2nd ed. London: E. & F. N. Spon, 1922.

Ramasami, Thirumalachari. "Approach Toward a Unified Theory for Tanning: Wilson's Dream," *Journal of the American Leather Chemists Association*, vol. 96, 2001, pp. 290-304.

Seligsberger, Ludwig. "Leather Research and Technology in the Age of Chrome," *Journal of the American Leather Chemists Association*, vol. 86, 1991, pp. 245-258.

Stellmach, Joseph J. "The Commercial Success of Chrome Tanning: A Study and Commemorative," *Journal of the American Leather Chemists Association*, vol. 85, 1990, pp. 407-424.

Wachsman, Hubert. "The Theory of Tanning," *World Leather*, April 2004, pp. 30-31.

Ward, A. G. "Collagen, 1891-1977: Retrospect and Prospect," *Journal of the Society of Leather Technologists and Chemists*, vol. 62, pp. 1-13.

Welch, Peter C. "A Craft That Resisted Change: American Tanning Practices to 1850," *Technology and Culture*, 1963, pp. 299-317.

——. *Tanning in the United States to 1850*. Washington, DC: Museum of History and Technology, Smithsonian Institution, 1964, pp. 2-29.

Leather Goods, Products, Craft

This section mostly excludes reference to shoes, which get their own section, in BORDER CROSSINGS, below.

Buirski, David, ed. *Sitting Comfortably: Upholstery Leathers Into the New Millennium*. Liverpool: World Trades Publishing, 1999.

De Recy, George. *The Decoration of Leather*. Translated by Maude Nathan. London: Archibald Constable & Co. Ltd., 1905.

Dowd, Anthony, compiler. *The Anthony Dowd Collection of Modern Bindings*. Manchester: The John Rylands University Library of Manchester, 2002.

Flanders, John. *The Craftsman's Way: Canadian Expressions*. Introduction, Hart Massey. Toronto: University of Toronto Press, 1981.

Hirschberg, Lynn. "In the Beginning, There Was Leather. . .". *New York Times Magazine*, November 30, 2003, pp. 114-115.

Hulme, E. Wyndham, et al. *Leather for Libraries*. London: Sound Leather Committee of the Library Association, 1905.

Hunter, George Leland. *Decorative Textiles*. Philadelphia: J. B. Lippincott Company, 1918.

Kanigel, Robert. "Made in USA—by Me," *The Leather Craftsman*, September 1987, p. 58.

Macgregor, Neil. *A Catalogue of Leather in Life, Art and Industry*. Northampton: Museum of Leathercraft, 1992.

The Market for Leather Goods in North America and Selected Western European Countries. Geneva: International Trade Centre, 1969.

Mobilio, Albert. "Genteel Readers of the World, Dig Deep," *Salon*, December 16, 1997.

Roth, Philip. *American Pastoral*. New York: Vintage, 1997.

Saddlemaking in Wyoming: History, Utility, Art. Catalog. Laramie: University of Wyoming Art Museum, 1993.

Taylor, Frederick W. "Notes on Belting," in *Two Papers on Scientific Management*. London: George Routledge & Sons, 1919.

Thomson, R. S. "Bookbinding Leather: Yesterday, Today and Perhaps Tomorrow," *Journal of the Society of Leather Technologists and Chemists*, vol. 85, pp. 66-71.

Waterer, John W. *Leather Craftsmanship*. New York: Frederick A. Praeger, 1968.

Willcox, Donald J., and James Scott Manning. *Leather*. Chicago: Henry Regnery Company, 1972.

Global Markets

The travails of the American and European leather industries, their losses to developing countries, and the fate of specific companies and industrial regions are not, quite, the subject of this book. But spend much time with tanners and they are never distant. Thanks to Mike Redwood for his recollections of Santa Croce in the 1970s, to my hosts in Ponte a Ebola, and to the American tanners listed above.

Amos, T. "Is the UK Leather Industry 'Finished'?" *Journal of the Society of Leather Technologists and Chemists*, vol. 85, pp. 199-202.
Blakey, R. "The Challenge of Change," *Journal of the Society of Leather Technologists and Chemists*, vol. 86, 2002, pp. 229-239.
Chan, Dominic S. "The Wonder of the Footwear Industry—China," *World Footwear*, July-August 2002, pp. 12-14.
"Deep-Seated Concerns for the Future," *World Leather*, October 2005, pp. 13-18.
"Italy—Fighting to Maintain Its Place in the World Market," *World Leather*, October 2004, pp. 17-35.
Koppany, E. John. "A Geopolitical Essay of the Leather Industry Over the Past 50 Years," *Journal of the American Leather Chemists Association*, vol. 99, no. 12, 2004, pp. 485-493.
"Pits, Aniline Dyes and Arsenic Paints?" *Leather International*, April 2002, pp. 53-54.
Redwood, Mike. "The Role of Marketing in Active Sportswear and Equipment," *Leather Industry Companion*, from a paper presented at the 37th International Man-Made Fibres Congress, Dornbirn, Austria, September 16-18, 1998.
——. "The Marketing of Leather and Leather Goods in Difficult and Changing Times," *Leather Industry Companion*, Winter 1998.
"Still Leaders but Times Are Hard," *Leather International*, May 2004, pp. 18-20.
"Still Leaving a Giant Footprint," *World Leather*, October 2002, pp. 27-35.

FAUX LEATHER

My story of the quest for artificial leather is told largely through four broad classes of them: (1) pyroxilin-based faux leathers such as Fabrikoid, (2) vinyl-based materials like Naugahyde, (3) Corfam and its "poromeric" cousins, and (4) microfiber-based synthetic leathers such as Ultrasuede. The bibliographic record, then, begins with these four classes and then turns to earlier synthetic leathers, to surveys of artificial leather generally, and to a few special trades and technologies, such as needlepunching and embossing, that figure in my story.

Pyroxilin

Correspondence and reports bearing on the early days of Fabrikoid, the pyroxilin-coated artificial leather featured here, and spanning the period from about 1908 to 1925, are largely drawn from the Hagley Museum and Library's archives. Some of these materials were originally made available to me through the courtesy of John Kenly Smith, Jr. Particularly notable are Experimental Station research reports, typically under the prefix B-95 and continuing over the course of some years, records from the federal government's antitrust action against Du Pont, and day-to-day correspondence bearing on Fabrikoid.

Thanks to Betsy McKean of the City of Newburgh's Records Management for letting me see plans of the Fabrikoid site, *circa* 1913; to Russell Lange of the Newburgh Historical Society and Chuck Thomas and Rita Forrester at the Newburgh Free Library for local articles bearing on Fabrikoid, for issues of *The Fabtonian*, the Du Pont Fabrikoid employee newsletter, and for the program of the "Fifth Annual Round Table Dinner of the Du Pont Fabrikoid Knockers' Club," December 28, 1915, which contains the lyrics of the immortal Fabrikoid Yell.

Articles, technical reports, and books bearing on pyroxilin-coated artificial leathers include the following:

Chase, Herbert. "How One Make of Artificial Leather Is Manufactured," *Automotive Industries*, vol. 49, August 2, 1923, pp. 224-227.

"Collier's Binds a Million Books in Fabrikoid," *Du Pont Magazine*, September-October 1921, pp. 4-5.

"Du Pont Advertising: Its Value to the Trade," *Du Pont Magazine*, March 1919.

Fabrikoid: An Improvement on Leather. Wilmington, DE: E. I. Du Pont de Nemours & Company, 1911.

Ginsberg, Ismar. "The Manufacture of Artificial Leather," *Scientific American Monthly*, October 1921, pp. 300-304.

Given, Guy Cumston. "Artificial Leather," *Industrial and Engineering Chemistry*, September 1926, vol. 18, no. 9, pp. 957-958.

Howell, William R. "The Manufacture of Du Pont Fabrikoid," *News-Letter*, Princeton Engineering Association, March 1929, pp. 67-69.

Marx, Carl. "Schoenbein, Discoverer of Cellulose Nitrate," *Plastics*, vol. 2, no. 1, January 1926, pp. 9+.

Meikle, Jeffrey L. "Presenting a New Material: From Imitation to Innovation with Fabrikoid," *Journal of the Decorative Arts Society*, vol. 19, 1995, pp. 8-15.

Neuberger, Rudolf. "History and Development of the Leather Cloth Industry," *Upholstering*, vol. 1, no. 4, July 1934, pp. 6+.

Patterson, J. R. "Bookbinding and the Newer Binding Materials." *Library Journal*, 1928.

Smith, John Kenly, Jr. "Fabrikoid—The Lesson of Leather," unpublished manuscript.

The Story of Du Pont Fabrikoid. Newburgh, NY: E. I. Du Pont de Nemours, 1931.

"Uncle Sam Says Not Enough," Du Pont American Industries advertisement. *The Tech* (Massachusetts Institute of Technology), August 14, 1917, p. 4.

Wescott, N. P. "How Coated Textiles Have Served," *Du Pont Magazine*, September 1927, pp. 28-29.

"When Lacquer and Fabric Meet," *Scientific American*, vol. 148, April 1933, pp. 228+.

Worden, Edward Chauncey. *Nitrocellulose Industry*. New York: D. Van Nostrand, 1911.

Vinyl

Naugahyde and other vinyl-based faux leathers are inescapably part of modern life, making for both an industrial story and a social one; I have tried to tell something of both. My account is developed in part from interviews with Michael S. Copeland, Martin Jacob, Grace Jeffers, Edward Nassimi, Jeff Post, Paul Wagner, and Bob Young. I am grateful to Jeff Post for arranging a visit to SanduskyAthol and to Ralph Maglio for his help in arranging my visit to the Crompton library, repository of Naugahyde and U.S. Rubber information, as well as to Crompton librarian Patricia Ann Harmon for her help while I was there, and to Randi Mates for her courtesy in sending me her Naugahyde files.

Articles and books bearing on polyvinyl chloride, vinyl-based imitation leathers, and Naugahyde itself include:

Kaufman, Morris. *The First Century of Plastics: Celluloid and Its Sequel.* London: The Plastics and Rubber Institute, 1963.

———. *The Chemistry and Industrial Production of Polyvinyl Chloride.* New York: Gordon and Breach, 1969.

Lois, George, with Bill Pitts. *What's the Big Idea?* New York: Plume, 1993, pp. 33-42.

Pitts, Bill, and George Lois. *The Art of Advertising: George Lois on Mass Communication.* New York: Harry N. Abrams, 1977.

The Research and Development Capability of the United States Rubber Company. United States Rubber Company, 1962.

Semon, Waldo Lonsbury. Internet biography. Online at: *http://www.bouncing-balls.com/timeline/people/s_semon.htm.*

Semon, Waldo L., and G. Allan Stahl. "History of Vinyl Chloride Polymers," BF Goodrich Research and Development Center, 1980.

The Story of US. United States Rubber Company, October 1948.

The Story of U.S. Naugahyde: The Finest in Plastic Upholstery. United States Rubber Company, ca. 1950.

Corfam

Despite Corfam's iconic status as an American business failure, little beyond the bare facts have appeared in book form until now. My account of its origins, rise, and fall has been built up from sources varied enough, I think, to warrant breaking down into the subcategories below. First, however, I wish to express my gratitude to the following men and women for their recollections of the Corfam project through in-person and telephone interviews: Libby Fay, Hamilton Fish, Dick Heckert, Ruth Ramsdell Holden, J. Lee Hollowell, John Korenko, Sam Lange, John Learnard, Thomas J. Leonard, Charles A. Lynch, Ron Moltenbrey, John Piccard, John C. Richards, Joe Rivers, and Bob Wilson. Special thanks to Mr. Moltenbrey and John Noble for extended email correspondence detailing aspects of Corfam production and, of course, to John and Mary Ann Piccard for their hospitality and goodwill.

Unpublished Sources

Archival sources, in-house Du Pont documents, and other unpublished papers from which my account draws include a great store of materials available at the Hagley Museum and Library, especially about early research on Corfam, before it was Corfam. These include extensive records, especially those covering the period 1949-1951, of the Pioneering Research Division.

The Hagley materials were substantially enriched by the deposit, not yet cataloged at the time of my visits there, of materials deposited by the late Joseph Leavy. My thanks to Marge McNinch for alerting me to them. They proved to be of inestimable value.

John Kenly Smith, Jr., was kind enough to let me see Corfam materials that he and David Hounshell gathered in researching their 1988 book, *Science and Corporate Strategy*.

John Piccard let me see his report "Non-woven Fabrics—I," August 30, 1949, a recounting of his earliest work with leatherlike materials.

Thanks to J. Lee Hollowell for access to his unpublished manuscript, "Horse Blankets to Haute Couture," October 22, 2001.

Betsy McKean, of the City of Newburgh's Records Management, arranged to get me maps of Du Pont's property in the late 1950s, when the Corfam project was gathering steam.

Among illuminating materials culled from these sources are the following:

"1966 'A' Bonus Recommendation," Du Pont Fabrics and Finishes Department typescript, March 29, 1967.
Adams, Carol. National Family Opinion, Inc., questionnaire. American Marketing Association, Toledo Chamber of Commerce.
Batson, H. E. "Report on Retailer Calls," Du Pont Fabrics and Finishes Department typescript, January 22, 1970.
Botsch, Richard C. "1970 Western Region Marketing Plan," Du Pont Poromeric Products Division typescript, January 6, 1970.
Burton, J. R. A. "Competition for Corfam," Du Pont typescript, January 1967.
"Corfam: A Research to Reality Case History," Du Pont typescript, ca. 1966.
"Corfam Technical Information," as defined in Article IV. a. and IV. 2. of the "Technical Information Sale Agreement" between E. I. du Pont de Nemours and Polimex-Cekop, Ltd., with respect to the production of poromeric materials in Poland, ca. 1972. [Hagley Acc. 1801].
Heckert, R. E. "Whither Corfam," Du Pont Fabrics and Finishes report to the Executive Committee, typescript, November 20, 1969.
Indoctrination and Training Manual, Dupont Poromeric Products Division, ca. 1968.
Lawson, W. D. Du Pont typescript of talk given at the press introduction of Corfam in New York, October 2, 1963.
——. "History and Analysis of Corfam," typescript, November 1972.
Leavy, J. B. "Corfam: Retail Sales Training," Du Pont Corfam Retail Marketing Bulletin, March 19, 1964.
——. Handwritten notes for Corfam presentations during the 1960s, including that at the Fashion Institute of Technology, November 4, 1968.
Lessing, Lawrence. "The Fast Footrace of Corfam," typescript, apparently commissioned by Du Pont, sent to editors as "a natural follow-up" to Lessing's 1964 article in *Fortune*, cited below, November 2, 1966.
Moyer, James E. "Corfam: An Advertising Case History." University of Illinois, College of Communications, ca. 1965. [Note on title page: "Prepared in collaboration with Du Pont personnel."]
Ogden, C. H. "1970 Territory Plan, Chicago-Wisconsin Territory," Du Pont Poromeric Products Division typescript.
"Question-Answer Fact Sheet: Corfam Poromeric Shoe Upper Material," Product Information Service, Du Pont Public Relations Department, January 17, 1964.
Rivers, Joe. Interview, conducted by David A. Hounshell, transcript, January 20, 1986.
Yuan, E. L. "The Structure and Property Relationships of Poromeric Materials," typescript, 1970.

William Rossi Collection

William Rossi was long-time executive editor of *Boot and Shoe Recorder*. From the William Rossi Collection, at Stonehill College, Easton, Massachusetts, numerous materials were made available through the courtesy of Nicole Tourangeau, archivist-librarian, and her work-study students, Sam

Gabrielson and Caragh McManus. Among the more notable of these are the following:

Condensation of talk given by William A. Rossi, Executive Editor, *Boot and Shoe Recorder*, at New England Tanners Production Club, typescript, Hawthorne Hotel, Salem, MA, January 17, 1964.

"Corfam: A Bright Star in Genesco's Future," in unknown Genesco company publication, February 1964.

"Corfam and Clarino Patent Pact Settles Conflict on Poromerics," *Footwear News*, March 6, 1969, p. 32.

Danzig, Fred. "Du Pont's Corfam: What Went Wrong?" *Advertising Age*, April 5, 1971, pp. 6+.

"Death Warrant for Corfam," *Footwear News*, March 18, 1971, pp. 1+.

"Poromerics, '60s Belle, Now Aging Spinster," *Footwear News*, October 28, 1971, pp. 1+.

"A Resume of the Presentation by E. I. Du Pont de Nemours & Co., Inc. to the International Shoe Company on Behalf of Corfam," typescript, ca. 1964.

Roddy, J. T. "U.S. Scientific Assessment of 'Corfam' Material Completed," *Leather*, May 22, 1964, p. 280.

——. "The Challenge of Synthetic Upper Materials," *Leather and Shoes*, April 1, 1967, pp. 40-45.

——. "The Case for Leather vs. Man-Made Materials," *Leather and Shoes*, April 20, 1968, pp. 16-23.

Technical Information on Shoemaking, Du Pont technical guide, October 12, 1965. Note in Rossi collection: "Prepared by W. A. Rossi for Du Pont—1963. For educational use by Corfam Division as introduction to basic shoe knowledge for Du Pont personnel."

Overviews

Over the years a number of broad-ranging reviews of the Corfam enterprise have appeared, including the following:

Carlson, Laurence Dale. "A Historical and Analytical Study of the New Product Introductions of a Man-Made Leather Shoe Bottom, Neolite, and Upper, Corfam." Dissertation, Ohio State University, 1967.

Jenkins, G. I., and H. G. Drinkwater. "Corfam Versus Cowhide: The Complete Case History," *The Director*, May 1969.

Lawson, William D., Charles A. Lynch, and John C. Richards. "Corfam: Research Brings Chemistry to Footwear," *Research Management*, vol. 8, no. 1, 1965, pp. 5-26.

Lessing, Lawrence. "Synthetics Ride Hell-Bent for Leather," *Fortune*, November 1964.

Littler, D. A., and A. W. Pearson. "Marketing a New Industrial Good: A Case Study," *Industrial Marketing Management*, vol. 3, 1972, pp. 299-307.

Pepper, K. W. "The Challenge of Corfam," The Director's Annual Lecture at the National Leathersellers College, February 18, 1965, published later in *Journal of the Society of Leather Trades*.

"The Story of 'Corfam': 25-year Journey from Dream to Reality," *Boot and Shoe Recorder*, October 1, 1963.

On the Eve of Corfam

The coming of Corfam represented a major crisis for the leather and footwear industries. My understanding of this key moment in industrial and commercial history was enriched by the following:

Boot and Shoe Recorder coverage, 1961-1964.
"Du Pont Plans for Synthetic Leather Output Heighten Rivalry for Huge Shoe Market," *Wall Street Journal*, December 27, 1962.
O'Flaherty, Fred. "Imitations and Substitutes," *Leather and Shoes*, September 24, 1960.
——. "The Invasion of the Tanning Industry," *The Leather Manufacturer*, December 1961.
"The Tanning Industry and Artificial Leather," confidential marketing research report, April 1962, prepared for the Dewey and Almy Chemical Company, including an extensive digest of trade articles and patent history bearing on artificial leather.

Technical Articles

Beck, P. J., and E. P. Lhuede. "The Case for Leather as a Shoe Upper Material," *Australian Leather Journal*, vol. 74, no. 11, March 1972, pp. 20-28.
Bossan, Louis Paul. "Advantages of 'Corfam' to the Shoe Manufacturer," *Rubber and Plastics Age*, February 1966, pp. 152-153.
Brooks, F. W., and R. G. Mitton. "Wear Trials for the Comparison of Leather and Synthetic Upper Materials in Shoes," *American Shoemaking*, October 11, 1967, pp. 8-19.
"Corfam—Leather-Substitute in Shoes Is Put to the Test of Use," *Consumer Bulletin*, vol. 48, no. 1, January 1965, pp. 25-26.
"Corfam vs. Leather for Shoe Uppers," *Consumer Reports*, November 1964, pp. 517-518.
Durst, Peter. "The PU Coagulation Process and Its Success in PUCF's," *Journal of Coated Fabrics*, vol. 13, January 1984, pp. 175-183.
Hole, L. G. "Poromerics: Their Structure and Use," *Rubber Journal*, April 1970, pp. 72-76.
Hole, L. G., and J. G. Butlin. "The Impact and Future of Man-Made Upper Materials," *Journal of the British Boot & Shoe Institution*, vol. 15, 1968, pp. 79-93.
Lawson, William D. "The Status of Man-Made Materials," *Boot and Shoe Recorder*, June 1, 1966.
Payne, A. R. *Poromerics in the Shoe Industry*. Amsterdam: Elsevier, 1970.
"Poromerics: How They're Manufactured," *Chemical Engineering News*, March 9, 1970, pp. 62-63.
Zorn, Bruno. "Porous Polyurethane Films and Coatings," *Journal of Coated Fabrics*, vol. 13, January 1984, pp. 166-173.

Other Publications

Anders, John. "This Cloth Has Been Fabricated," *The Dallas Morning News*, January 20, 1988.
"Another Nylon," *Forbes*, October 15, 1964, pp. 15-16.
Barnfather, Maurice. "Polish Joke," *Forbes*, March 2, 1981, p. 46.
Cortz, Dan. "The $100-Million Object Lesson," sidebar to "Bringing the Laboratory Down to Earth," *Fortune*, January 1971.
Culberson, Fred Ray. "The Corfam Failure." Dissertation, University of Texas at Austin, 1971.
Davis, Harry E. "'Corfam': First Focus Is Footwear," *Du Pont Magazine*, November-December 1963, pp. 2-5.
"Du Pont Does It," *Forbes*, December 15, 1969, pp. 22-24.
"Exit Corfam," *Barron's National Business and Financial Weekly*, March 22, 1971, p. 1.
"Fiber Fact Finders," *Du Pont Magazine*, January-February 1961, pp. 19-21.
Flanigan, James J. "Stepping Ahead with Corfam," *New York Herald Tribune*, December 8, 1963.
"Getting Corfamiliar," *Leather and Shoes*, vol. 146, no. 15, October 12, 1963.
Hemp, Paul. "Free the Wrinkle," *Boston Globe*, September 11, 1994.
Lawson, W. D. "Status of Corfam," *Boot and Shoe Recorder*, vol. 171, July 1967, p. 51.
May, Roger B. "Du Pont Faces a Race as Big Competitors Take a Shine to Synthetic-Leather Market," *The Wall Street Journal*, September 19, 1966, p. 32.
McCormick, James H. "When Do You Drop a Product?" *Du Pont Magazine*, June-July 1957, pp. 14-15.
"New Laboratory Issue." *The Fabtonian*, July 1948.
Pepper, K. W. "The Challenge of Synthetics to Leather," *Chemistry and Industry*, December 10, 1966, pp. 2079-2085.
"Reflections on the Reader's Digest Case," *Leather and Shoes*, vol. 147, no. 13, March 28, 1964, p. 4. See also: Don Wharton. "Nylon—a Triumph of Research," *Textile World*, January 1940, pp. 50-52; "Big News in Shoes," *Readers Digest*, March 1964; and brief biography of Wharton in Don Wharton Papers, University of North Carolina at Chapel Hill, Southern Historical Collection.
"Research: If the Shoe Fits, Another Winner for Industry," *Newsweek*, April 6, 1964.
Robertson, Andrew. "How Du Pont's Corfam Took a $100m Tanning," *Sunday Times*, March 21, 1971.
Sloane, Leonard. "Advertising: Bout with Manmade 'Leather'," *New York Times*, July 26, 1964, p. F12.
——. "Du Pont's $100-Million Edsel," *New York Times*, April 11, 1971, p. F3.
"Synthetics: Good Fit in Footwear," *Chemical Week*, May 1, 1965, p. 45.
"The Withdrawal of Corfam," from "Poromerics Progress," *SATRA*, vol. 3, no. 2, April 1971, pp. 77-82.

Microfiber

I am extraordinarily grateful to Miyoshi Okamoto, whose recollections, insights, and technical clarifications of his early work on Ultrasuede, through a months-long email correspondence, were crucial to my story.

Mr. Okamoto was endlessly patient in responding to my queries in English. At one point he resorted to making a stick-figure drawing of a key early experiment; it has found a permanent place, suitably framed, on my wall. Contributing also to my understanding of Mr. Okamoto's work was his Japanese-language research memoir, cited below, expertly translated from the Japanese by Homer Reid. Thanks to Ian Condry and Yoshimi Nagaya for helping me arrange the translation; and to Greg Ornatowski and Carl Accardo for their insights into Japanese corporate life.

I met with Atsushi Kumano and Sachiko Barbasso of Kuraray and Yoshifumi Hamada, Yasuhiro Takagi, and Toshinori Hara of Toray. I am enormously grateful, too, to Mira Zivcovich for the glimpse into the creative process that her work with Amaretta afforded me.

My account of Lorica is based in part on a tour of the Lorica factory in Sardinia and interviews with Giorgio Savini, Giuseppe Munafò, and Enrico Racheli over a two-day period in March 2005; Signor Racheli translated the comments and explanations of his colleagues.

My account profits as well from Amanda Lindsay's doctoral dissertation devoted to microfibers, cited below, and the useful email correspondence it stimulated.

Among articles and books that helped me tell the story of microfiber-based artificial leathers are these:

Ajgaonkar, D. B. "Microfibres," *Man-Made Textiles in India*, September 1992, pp. 327-337.

Dullea, Georgia. "Machine-Washable 'Suede,'" *New York Times*, March 23, 1976.

Hoashi, Koji. "Suede-Type Man-Made Leather for Clothing," *Japan Textile News*, no. 269, April 1977, pp. 92-95.

Hongu, Tatsuya, and Glyn O. Phillips. *New Fibers*, 2nd ed. Cambridge: Woodhead Publishing, 1997.

Ivins, Molly. "The Fabrics That Define Republican Women," *Dallas Times Herald*, August 27, 1984.

Japan Business History Institute, ed. *The History of Toray 70: 1926-1996*. Tokyo: Toray Industries, 1999.

Lindsay, Amanda. "The Evolution of Microfibre Through Technology and Market Pressure." Dissertation, University of Sussex, 1999.

——. "Product and Process Innovation in the Chemical Fibre Industry: Patenting in Microfibres." Unpublished manuscript. London Metropolitan University, London, August 17, 2002.

May-Plumlee, Traci, and Thomas F. Gilmore. "Ultrasuede: Nonwoven Technology—Lessons from Nature." *International Nonwovens Journal*, vol. 7, no. 3, 1995, pp. 39-48.

——. "Ultrasuede: Nonwovens Imitate Nature." Papers of INDA-TEC 95. Association of the Nonwoven Fabrics Industry, 1995, pp. 237-255.

Mukhopadbyay, Samrat, "Microfibres—an Overview," *Indian Journal of Fibre & Textile Research*, vol. 27, September 2002, pp. 307-314.
Nakajima, T. *Advanced Fiber Spinning Technology*, English edition, K. Kajiwara and J. E. McIntyre, eds. Cambridge: Woodhead Publishing, 1994.
Neff, Robert. "Toray May Have Found the Formula for Luck," *Business Week*, June 25, 1990, p. 57.
Okamoto, M. "Ultra-fine Fiber and Its Application," *Japan Textile News*, two parts. November 1977, pp. 94-97, and January 1978, pp. 77-81.
——. "Ultra-Fine Fibres: A New Dimension for Polyester," in *Polyester: 50 Years of Achievement*, David Brunnschweiler and John Hearle, eds. Manchester: The Textile Institute, 1993, pp. 108-111.
——. "An Unprecedented New Suede-like Material: A Research Memoir," *Gubrafar-to-ki* [Expected Materials for the Future], two parts, vol. 5, nos. 1 and 2, 2005. Translated by Homer Reid.
Okamoto, M., and K. Kajiwara. *Shingosen: Past, Present, and Future.* Technomic Publishing Co., 1970.
Robertson, James, and Michael Grieve, eds. *Forensic Examination of Fibres*, 2nd ed. London: Taylor and Francis, 1999, pp. 408-419.
Wedemeyer, Dee. "Ultra Demand for Versatile Ultrasuede," *New York Times*, February 26, 1977.

Early Imitation Leathers

For background on Japanese leather paper, I am grateful to Yasuko Suga, Tsuda College, Tokyo, and to Greg Herringshaw of the Cooper-Hewitt Museum, New York, for showing me samples of both Japanese leather paper and the Spanish leather it sought to imitate.

The following books and articles enriched my understanding of early faux leathers, including Japanese leather paper:

Barrett, Timothy. *Japanese Papermaking: Traditions, Tools, and Techniques.* New York: Weatherhill, 1983.
Christie, Guy. *Storeys of Lancaster, 1848-1964.* London: Collins, 1964.
Hall, J. Sparkes. *The Book of the Foot: A History of Boots and Shoes.* New York: William H. Graham, 1847.
Hughes, Sukey. *The World of Japanese Paper*, Tokyo: Kodansha International, 1978.
Leatherette [Harrington & Co.'s New Substitute for Leather]. London: Richards, White & Co., 1875.
Madru, René. "Quelques notes sur les cuirs artificiels," *Collegium*, May 1913, pp. 209-213.
Rein, J. J. *The Industries of Japan.* New York: A. C. Armstrong, 1889.
Thorp, Valerie. "Imitation Leather: Structure, Composition and Conservation," *Leather Conservation Newsletter*, vol. 6, no. 2, Spring 1990, pp. 7-15.
Watererer, John W. *Spanish Leather.* London: Faber and Faber, 1971.

Yokoyama, Yuko. "Takashi Ueda's Kinkarakami," March 2003. Online at: *http://www.handmadejapan.com/e_/index_e.html*.

Surveys and Reviews

In the list below, I've aimed to include all the more useful surveys and reviews of artificial leathers I've encountered:

"Artificial Leathers—Their Manufacture, Properties, and Uses," *Proceedings of the First SATRA International Conference*, Blackpool, England, 1971.

Buirski, David. "Genuine or Not—It's Here to Stay," *World Leather*, October 2002, pp. 89-93.

Civardi, F. P., and G. R. Hutter. "Leatherlike Materials," *Encyclopedia of Chemical Technology*, 3rd ed., vol. 14. New York: John Wiley & Sons, 1978-1984, pp. 231-249.

"The Evolution of Synthetics for the Shoe Industry," *Journal of Coated Fabrics*, vol. 6, January 1977, pp. 176-181.

Hayashi, Takafumi. "Man-Made Leather," *Chemtech*, January 1975, pp. 28-33.

Hioki, Katsumi, "Leather-like Materials," *Kirk-Othmer Encyclopedia of Chemical Technology*, 4th ed., vol. 15. New York: John Wiley & Sons, 1991-1998.

Hole, L. Geoffrey. "Artificial Leathers," *Reports on the Progress of Applied Chemistry During 1972*, vol. 57, 1973, pp. 181-206.

Hollowell, J. L. "Leather-like Materials," *Encyclopedia of Polymer Science and Technology*, vol. 8. New York: Interscience Publisher, 1964-1972, pp. 210-231.

Kruse, Hans-Hinrich, and J. H. Benecke. "Leather Imitates," *Ullmann's Encyclopedia of Industrial Chemistry*, 5th ed. Buchholz: VCH, 1990.

List of United States, British and German Patents Covering the Manufacture of Leather Substitutes, compiled by Mock & Blum, patent lawyers, 1918.

Nagoshi, Kazuo. "Clarino, Man-Made Leather," *International Progress in Urethanes*, vol. 3, 1981, pp. 193-217.

Nagoshi, K. "Leatherlike Materials," *Encyclopedia of Polymer Science and Engineering*, 2nd ed., vol. 8. New York: John Wiley & Sons, 1985-1989.

Payne, A. R. "Trends in Poromerics and Coated Fabrics in the Footwear and Allied Industries," presented at the Fifth Shirley International Seminar on the Place of Textiles in the Economy of a Developed Country. Manchester, England: Shirley Institute, 1972.

Sittig, Marshall. *Synthetic Leather from Petroleum*. Park Ridge, NJ: Noyes Development Corporation, 1969.

Suskind, Stuart P. "Man-Made Leather Substrates," *Journal of Coated Fibrous Materials*, vol. 2, April 1973, pp. 187-195.

Swedberg, Jamie. "The Truth About Faux Leather," *Industrial Fabric Products Review*, April 1999, pp. 26-30.

Whittaker, R. E. "Structure and Viscoelastic Properties of Poromerics," *Journal of Coated Fibrous Materials*," vol. 2, July 1972, pp. 3-23.

Trades and Technologies

My understanding of the making of embossing rollers, used for leather both faux and real, was enriched by a visit to the facilities of Standex Engraving, Sandston, Virginia. There Max Roelen, Gerd Mirtschin, Tommy Austin, and others introduced me to the field. Mr. Mirtschin, especially, took time both during my visit and in long phone conversations afterward to convey something of its intricacies.

What follows are books and articles, beyond those cited above, on some of the trades and technologies that figure in the making of artificial leathers:

Albrecbt, Wilheml, Hilmar Fuchs, and Walter Kittlemann, eds. *Nonwoven Fabrics* Weinheim: Wiley-VCf1, 2003.

Durst, Peter. "PU Transfer Coating of Fabrics for Leather-like Fashion Products," *Journal of Coated Fabrics, vol. 14,* April 1985, pp. 227-241.

Edwards, Kenneth N. "A History of Polyurethanes," in *Organic Coatings Their Origin and Development, R.* B. Seymour and 11. E Mark, eds. Amsterdam: Elsevier, 1990.

"Embossing Leather Substitutes," *Du Pont Magazine,* February 1919, pp. 26-27.

Lomax, Robert. "Recent Developments in Coated Apparel," Journal *of Coated Fabrics, vol.* 14, October 1984, pp.)1-9).

Liinenschloss, J., and W Chichester Albrecht, eds. *Non-Woven Bonded Fabrics.* New York: Halsted Press, 1985.

The Nonwovens Handbook. New York: INDA, Association of the Nonwoven Fabrics Industry, 1988.

Oertel, Gunter, ed. Polyurethane *Handbook,* 2nd ed. Cincinnati: Hanser Gardner Publications, 1993.

Schore, Elias. "Electroforming, Embossing, and Graining plates," *Metal Finishing,* January 1954, pp. 74-76.

Smith, Philip A. '"The Technology of Non-Woven Fabrics for Artificial Leathers," in "Artificial Leathers-Their Manufacture, properties, and Uses," *Proceedings of Me First SATRA International Conference,* Blackpool, England, 1971, pp. 107-115.

Vaughn, Edward A. "Historic Needlepunch Developments," *Nonwovens Industry,* March 1992, pp. 44-46.

BORDER CROSSINGS

This book necessarily straddles, explores, and probes the borders between natural and synthetic, real and fake. What follows is a compilation of bibliographic and other sources for some of the disparate fields into which my research has taken me.

Shoes

Leather has always been linked to shoes, and the footwear industry has been a key battleground between leather and synthetics, especially Corfam. My appreciation for shoes, as both utilitarian and aesthetic objects, and my understanding of shoe materials and of foot comfort, owes much to Mike Redwood and Jack Erickson for graciously arranging a tour of the Foot Joy shoe factory, Brockton, MA; to John Learnard for a tour of the Brockton Shoe Museum; to Megan Ogilvie, my former student in MIT's Graduate Program in Science Writing, for notes of her visit to the Bata Shoe Museum, Toronto; and to countless footwear industry salespeople, including members of the Boston Shoe Travelers Association, for their insights and recollections.

Listed below are books, articles, and archival records, some devoted to Corfam's challenge to leather, that proved especially useful. Unpublished documents listed below can be found at the Hagley Museum and Library, Wilmington, DE.

"2,000 Miles to Minneapolis," *Du Pont Magazine*, vol. 59, no. 6, November-December 1965.

Bilger, Burkhard. "Sole Survivor," *The New Yorker*, February 14 and 21, 2005.

Bitlisli, B. O., et al. "Importance of Using Genuine Leather in Shoe Production in Terms of Foot Comfort," *Journal of the Society of Leather Technologists and Chemists*, vol. 89, pp. 107-110.

Brooks, F. W., and R. G. Mitton. "Wear Trials for the Comparison of Leather and Synthetic Upper Materials in Shoes," *American Shoemaker*, October 11, 1967, pp. 8+.

Burton, J. R. A. "Physiological and Hygienical Aspects of the Use of Corfam Poromeric Material in Footwear." Draft of presentation intended for the German Medical Association, according to Du Pont cover letter, August 2, 1967.

——. "Corfam—Foot Comfort and Health." Du Pont typescript, May 1, 1968.

Carey, Bill. "King Maxey," chapter in *Fortunes, Fiddles & Fried Chicken: A Nashville Business History*. Franklin, TN: Hillsboro Press, 2000.

Cohen, Richard L., ed. *The Footwear Industry: Profiles in Leadership*. New York: Fairchild Publications, 1967.

Cohn, Walter E. *Modern Footwear Materials & Processes*. New York: Fairchild Publications, 1969.

DeLano, Sharon, and David Rieff. *Texas Boots*. New York: Viking Press, 1981.

Diebschlag, Wilfried, and Wolfgang Nocker. "A Comparative Analysis of the Comfort of Leather and Substitute Materials, Especially for Footwear," *Journal of the American Leather Chemists Association*, vol. 73, 1978, pp. 307-332.

Diebschlag, Wilfried, et al. "The Influence of Several Socks and Linings on the Microclimate in Shoes with Upper Material of Leather or Synthetic," *Journal of the American Leather Chemists Association*, vol. 71, no. 6, June 1976, pp. 293-306.

"Footwear Fundamentals," parts 1-8, various authors, *World Footwear*, March-April 2002 to May-June 2003.
Garley, Tony, "Leather Cutting: Yield Versus Quality," *World Footwear*, September-October 2002, pp. 33-34.
Heath, Arthur L. "Nurses' Service Shoe Wear Test," Du Pont typescript, November 6, 1967.
Hill, L. M., and S. G. Shuttleworth. "Comfort Factors in Shoe Upper Materials," *Journal of the American Leather Chemists Association*, 1971, pp. 5-20.
Hole, L. G., and B. Keech. "Foot Comfort Properties of Natural and Artificial Leathers," in "Artificial Leathers—Their Manufacture, Properties, and Uses," *Proceedings of the First SATRA International Conference*, Blackpool, England, 1971, pp. 405-424.
Hoover, Edgar M., Jr. *Location Theory and the Shoe and Leather Industries*. Cambridge: Harvard University Press, 1937.
Kennedy, J. E. "A Study of the Microclimate in Footwear," *Journal of the American Leather Chemists Association*, vol. 62, no. 5, May 1967, pp. 310-333.
Maeser, Mieth. "An Engineer Looks at Leather," *Journal of the American Leather Chemists Association*, vol. 58, August 1963, pp. 456-493.
Meldman, Edward C., and Arthur E. Helfand. "Compatibility of du Pont Corfam with Efficient Foot Function," *Journal of the American Podiatry Association*, May 1965, pp. 351-355.
O'Flaherty, Fred. "Technical Developments in Shoe Leathers," *The Leather Manufacturer*, April 1961.
——. "Leather Breathing: Contribution to Foot Comfort," *Leather & Shoes*, January 25, 1963.
O'Keefe, Linda. *Shoes: A Celebration of Pumps, Sandals, Slippers & More*. New York: Workman Publishing, 1996.
Redwood, Mike. "The Demands Made on Leather by New Processes for the Manufacture of Footwear and Other Goods," *Leather Industry Companion*, Donald Burton Prize Essay, 1969.
Swann, June. *Shoes*. New York: Drama Book Publishers, 1982.
Vass, László, and Magda Molnár. *Handmade Shoes for Men*. Cologne: Könemann, 1999.
Venkatappaiah, B. *Introduction to the Modern Footwear Technology*. Chennai: Central Leather Research Institute, 1997.
Vickers, Robert A., and Fred O'Flaherty. "Transpiration and Other Inherent Properties of Leather," reprinted from *The Leather Manufacturer*, ca. 1950.

Car Seats

Another battleground for leather and artificial leather has been automotive upholstery, a subject featured at the 100th anniversary meeting of the American Leather Chemists Association in St. Louis in 2004 and recounted in issues of the association's journal in subsequent months.

"Heated Seats: Getting Too Hot," *World Leather*, April 2005, pp. 23-26.
Hur Yoon-Sook and Se-Jin Park. "Evaluation of Comfort Properties with Covering Textiles of Car Seats," *Human Factors in Driving, Vehicle Seating, and Rear Vision, Proceedings of the 1998 SAW International Congress & Exposition*, February 23-26, Detroit, MI, pp. 63-70.

Phelan, Mark. "Interiors Slip Into Something Leather," *Automotive Industries*, May 1998.
Tanaka, N., and J. Tanaka. "New Man-Made Leather for the Automotive Industry," *Melliand International*, vol. 6, June 2000, pp. 137-138.
Taub, Bernard. "Urethane Coated Fabrics for Automotive Upholstery Applications," *Journal of Coated Fibrous Materials*, vol. 2, January 1973, pp. 135-146.
"Testing Automotive Leather," *Leather International*, May 2003, 2 pp.
Winter, Drew. "A Second Look at Vinyl," *Ward's Auto World*, July 1, 1995.

Other Commercial Battlegrounds

"Agricultural Raw Materials: Competition with Synthetic Substitutes," Commodities and Trade Division, Food and Agriculture Organization of the United Nations, Rome 1984.
Booth, Hannah. "Great Comfort," *Design Week*, September 5, 2002, pp. 27-28.
"Impact of Synthetics on Markets for Natural and Traditional Materials," European Association for Industrial Marketing Research ECMRA Conference, Aix-en-Provence, 1973.
Lee, W. K. "Textile Leather and Poromerics," in two parts, *Textile Asia*, March and April 1977, 8 and 6 pp, respectively.
Lunden, Bo, and Ference Schmel. "Soft Leather Substitute Materials and Their Impact on the International Leather and Leather Products Trade," *Journal of Coated Fabrics*, vol. 14, July 1984, pp. 9-35.
McCaffety, Cynthia Reece. "Speaking Volumes," *Hemispheres*, September 2004, pp. 56-60.
Morley, Derek. "Coated Fabrics for Upholstery," *Journal of Coated Fabrics*, vol. 14, July 1984, pp. 46-52.
Saada, Michael, and John McWethy. "Man-Made Leather: Synthetic Battles Way Into More Billfolds, Shoes and Suitcases," *The Wall Street Journal*, October 22, 1951.
Thon, Bernard. "The Plastic Saddle," *Western Horseman*, vol. 51, no. 1, pp. 6-7.
Wilstein, Steve. "Leather Balls Go Way of Wooden Rackets and Bats," Associated Press, May 16, 2002.

Fashion and Popular Culture

Leather and its imitators hold a key place in fashion and popular culture, sometimes figuring in controversy bearing on animal rights, sexuality, and other issues. For their insights on such matters, I am grateful to the following for granting me interviews: Michele Bryant, Francesca Sterlacci, Valerie Steele, Grace Jeffers, and Erika and Sarah Kubersky. Thanks to Jose Madera for letting me sit in on his leather apparel class at the Fashion Institute of Technology; to Jennifer Cohlman at the Cooper-Hewitt design museum in New York; to Sara J. Kadolph, textile and clothing program, Iowa State University; to Shelly Foote at the Smithsonian's Division of Social History; to Edward Turk, literature professor at MIT; and to the many people I talked to at the New England Fetish Fair. The public stand of People for the Ethical Treatment of Animals is voiced, formidably, through *Animal Times* and the group's many other publications.

Bronski, Michael. *The Pleasure Principle.* New York: St. Martin's Press, 1998, pp. 88-108.
Dichter, Ernest. *Handbook of Consumer Motivations.* New York: McGraw-Hill, 1964.
——. *Motivating Human Behavior.* New York: McGraw-Hill, 1971.
Foster, Vanda. *Bags and Purses.* New York: Drama Book Publishers, 1982.
Gottlieb, Robert, and Frank Maresca, eds. *A Certain Style: The Art of the Plastic Handbag, 1949-59.* New York: Alfred A. Knopf, 1988.
Gross, Elaine, and Fred Rottman. *Halston: An American Original.* New York: HarperCollins, 1999.
Hass, Nancy. "Losing the Fur War, PETA Advances Attack on Leather," *New York Times,* March 23, 2000.
Hine, Thomas. *Populuxe.* New York: MJF Books, 1999.
Johnson, Anna. *Handbags: The Power of the Purse.* New York: Workman Publishing, 2002.
Lurie, Alison. "Fashion and Sex," in *The Language of Clothes.* New York: Random House, 1981, pp. 230-234.
Mohr, Richard D. *Gay Ideas: Outing and Other Controversies.* Boston: Beacon Press, 1992.
Postrel, Virginia. *The Substance of Style: How the Rise of Aesthetic Value Is Remaking Commerce, Culture, and Consciousness.* New York: HarperCollins, 2003.
Singer, Peter. *Animal Liberation: A New Ethics for Our Treatment of Animals.* New York: Avon Books, 1975.
Specter, Michael. "The Extremist," *The New Yorker,* April 14, 2003, pp. 52+.
Steele, Valerie. *Fetish: Fashion, Sex and Power.* New York: Oxford University Press, 1966.
Steele, Valerie, and Laird Borrelli. *Handbags: A Lexicon of Style.* New York: Rizzoli, 1999.
Wilcox, Claire. *A Century of Bags: Icons of Style in the 20th Century.* New York: Chartwell Books, 1997.
Williams, Rosalind. *Dream Worlds: Mass Consumption in Late Nineteenth-Century France.* Berkeley: University of California Press, 1982.

Human Senses

Synthetics have sought to emulate leather not just functionally, but aesthetically and sensually as well. Thanks to Mandayam Srinivasan, Chris Moore, and Seshadri Ramkumar for their clear explanations of the sense of touch; to Stephen Warrenburg and Guillermo Fernandez for drawing me into the world of smell; and to Willem van Eijk for coordinating my visit to International Flavors and Fragrances, Hazlet, NJ.

Alexander, K. T. W., and R. G. Stosic. "A New Non-Destructive Leather Softness Test," *Journal of the Society of Leather Technologists and Chemists,* vol. 77, March 1993, pp. 139-142.
Bishop, D. P. "Fabrics: Sensory and Mechanical Properties," *Textile Progress,* vol. 26, no. 3, pp. 1-64.

Hancock, Elise. "A Primer on Touch," *Johns Hopkins Magazine*, September 1996.

Kleban, Martin. "Leather with an Aroma," *World Leather*, April 2002, pp. 69-70.

Landman, W. W., R. G. Stosic, J. Vaculik, and M. Hanson. "Softness—an International Comparison," *Journal of the Society of Leather Technologists and Chemists*, vol. 78, January 1994, pp. 88-92.

Long, A. J., et al. "The Use of Acoustic Emission as an Aid to Evaluating the Handle of Leather," *Journal of the Society of Leather Technologies and Chemists*, vol. 85, no. 5, September-October 2001, pp. 159-163.

Mathes, Sharon, and Kay Flatten. "Performance Characteristics and Accuracy in Perceptual Discrimination of Leather and Synthetic Basketballs," *Perceptual and Motor Skills*, vol. 55, 1982, pp. 128-130.

Pye, David. *The Nature and Art of Workmanship*. Bethel, CT: Cambium Press, 1995.

Su Zhenwei, Yin Guofu, and Zhuo Zhaofei. "Objective Evaluation of Leather Handle with Artificial Neural Networks," *Journal of the Society of Leather Technologists and Chemists*, vol. 80, November 1995, pp. 106-109.

Troy, D. J. "The Appearance of Poromeric Materials," in "Artificial Leathers—Their Manufacture, Properties, and Uses," *Proceedings of the First SATRA International Conference*, Blackpool, England, 1971, pp. 426-444.

Plastics and Polymers

Barthes, R. "Plastic," in *The Everyday Life Reader*. B. Highmore, ed. London: Routledge, 2002.

Brooke, Walter. Interview. *Fresh Air*, NPR, January 8, 2003.

Fenichell, Stephen. *Plastic: The Making of a Synthetic Century*. New York: HarperCollins Publishers, 1996.

Friedel, Robert. *Pioneer Plastic: The Making and Selling of Celluloid*. Madison: University of Wisconsin Press, 1983, pp. 59-89.

Furukawa, Yasu. *Inventing Polymer Science: Staudinger, Carothers, and the Emergence of Macromolecular Chemistry*. Philadelphia: University of Pennsylvania Press, 1998.

Hermes, Matthew. *Enough for One Lifetime: Wallace Carothers, Inventor of Nylon*. Washington, DC: American Chemical Society and the Chemical Heritage Foundation, 1996.

Mark, H. "Coming to an Age of Polymers in Science and Technology," *History of Polymer Science and Technology*, R. B. Seymour, ed. New York: Marcel Dekker, 1982.

McKie, Douglas. "Wöhler's 'Synthetic' Urea and the Rejection of Vitalism: A Chemical Legend," *Nature*, vol. 153, May 20, 1944, pp. 608-610.

Meikle, Jeffrey L. *American Plastic: A Cultural History*. New Brunswick, NJ: Rutgers University Press, 1995.

Nunberg, Geoffrey. "Plastic: The Word Has Gone Undercover," *The Mercury News*, January 5, 2003.

Slack, Charles. *Noble Obsession*. New York: Hyperion, 2002.

Staudinger, Hermann. "Macromolecular Chemistry," Nobel Lecture, December 11, 1953.

Wöhler, F. "On the Artificial Production of Urea," *Annalen der Physik und Chemie*. Leipzig, 1828.

Central to virtually any account of synthetic materials is E. I. Du Pont de Nemours, two of whose products, Fabrikoid and Corfam, figure prominently in my story. Du Pont is among the most studied of companies and has accumulated its own thick bibliographical record. My account draws often from Hounshell and Smith's important work, cited below; special thanks to John Kenly Smith for granting me access to his papers and to his unpublished account of Fabrikoid. Thanks to Deb Liczwek for her tour of the Experimental Station; to Trudy Batelic who showed me around the old Du Pont research lab in Newburgh, NY; and again to Marge McNinch at the Hagley Museum and Library. The website *http://heritage.dupont.com* carries many useful references to Du Pont history.

Carey, Bill. "Gunpowder & Fighter Planes," in *Fortunes, Fiddles & Fried Chicken: A Nashville Business History*. Franklin, TN: Hillsboro Press, 2000.

Du Pont: The Autobiography of an American Enterprise. Wilmington, DE: E. I. Du Pont de Nemours & Co., 1952.

du Pont de Nemours, E. I. *Our Old Hickory Heritage*. Wilmington, DE: E. I. Du Pont de Nemours & Co., 1982.

Hounshell, David A., and John Kenly Smith, Jr. *Science and Corporate Strategy: Du Pont R&D, 1902-1980*. Cambridge: Cambridge University Press, 1988.

Kinnane, Adrian. *Du Pont: From the Banks of the Brandywine to Miracles of Science*. Baltimore: Johns Hopkins University Press, 2002.

"The Master Technicians." *Time*. November 27, 1964, 5 pp.

Neumann, Laura Diann. "Strategic Marketing of High Price, High Quality Fashion Products: A Du Pont Case Study." Dissertation, University of Missouri-Columbia, 1993.

Materials

Edwards, Clive. *Encyclopedia of Furniture Materials, Trades and Techniques*. Ashgate Publishing, 2000.

Lupton, Ellen. *Skin: Surface, Substance and Design*. New York: Princeton Architectural Press, 2002.

Menzel, Peter. *Material World: A Global Family Portrait*. San Francisco: Sierra Club Books, 1994.

Okamoto, Miyoshi. "Polymer Materials Which Appeal to Kansei," in *Progress in Pacific Polymer Science* 2, Y. Imanishi, ed. Berlin: Springer-Verlag, 1992, pp. 345-353.

Patton, Phil. "A Wealth of Materials That Say 'Material Wealth,'" *New York Times*, May 9, 2005, p. D8.

Seelig, Warren. "Thinking Aloud: Contemporary Fiber, Material Meaning," *American Craft*, vol. 65, no. 4, August-September 2005, pp. 42+.

Simpson, Pamela H. *Cheap, Quick, & Easy*. Knoxville: University of Tennessee Press, 1999.

Faux and Real

A conference at the Dibner Institute for the History of Science and Technology, Cambridge, MA, on May 18-19, 2001, "The Artificial and the Natural: An Ancient Debate and Its Modern Descendants," organized by William Newman and Berndadette Bensaude-Vincent, offered me early entrée to the endlessly fascinating terrain of natural and synthetic, imitation and copy, real and fake. Herewith is an assortment of sources and provocations:

Alsberg, Carl L. "Economic Aspects of Adulteration and Imitation," *Quarterly Journal of Economics*, vol. 46, no. 1, December 1931, pp. 1-33.

April Fool: Folk Art Fakes and Forgeries. Catalog of exhibition at Museum of American Folk Art, New York, April 1-30, 1988.

Barash, David P. "Nature Takes Only Tiny Steps but Still Surpasses Our Reckoning," *The Chronicle of Higher Education*, April 18, 2003.

——. "The Tyranny of the Natural," *The Chronicle of Higher Education*, November 2, 2001.

Bartindale, Becky. "The Perfect Tree?" *Mercury News*, November 27, 2005.

Barton, Laura. "Flight from Reality," *The Guardian*, August 16, 2003. Online at: *http://www.guardian.co.uk.*

Baudrillard, Jean. *Simulacra and Simulation.* Translated by Sheila Faria Glaser. Ann Arbor: University of Michigan Press, 1994.

Benjamin, Walter. "The Work of Art in the Age of Mechanical Reproduction," in *Illuminations*, Hannah Arendt, ed., translated by Harry Zohn. New York: Schocken Books, 1969.

Berg, Maxine. "From Imitation to Invention: Creating Commodities in Eighteenth-Century Britain," *Economic History Review*, vol. LV, no. 1, 2000, pp. 1-30.

Boyle, David. *Authenticity: Brands, Fakes, Spin and the Lust for Real Life.* London: Flamingo, 2003.

Coleman, David. "Reality Check—Imitation Is the Sincerest Form of Flattery, Not the Chicest," *Vogue*, May 2001.

Dean, Irene Semanchuk. *Faux Surfaces in Polymer Clay: 30 Techniques That Imitate Precious Stones, Metals, Wood & More.* New York: Lark Books, 2003.

Des Jardins, Andrea. "Determining the 'Naturalness' of a Product." Online at: *http://www.herc.org.*

Garfield, Simon. *Mauve: How One Man Invented a Color That Changed the World.* New York: W.W. Norton, 2000.

Gayford, Martin. "It Makes You Think," *The Spectator*, December 29, 2001.

Gorman, Barbara. "Women's Attitudes Toward Simulated Furs," Du Pont typescript report, August 17, 1976. Hagley Museum.

Huysmans, Joris-Karl. *Against Nature*, translated by Robert Baldick. New York: Penguin, 1959.

Leland, John. "Beyond File-Sharing, A Nation of Copiers," *New York Times*, September 14, 2003.

Lewis, Michael J. "It Depends on How You Define 'Real'," *New York Times*, June 23, 2002.

Marx, Leo. *The Machine in the Garden.* New York: Oxford University Press, 2000. Marx conceives a "middle landscape" between the natural and human-built worlds. Leather—rooted in nature, transformed by art—would seem a particularly comfortable inhabitant in it.

Newman, Morris. "A Different Sort of Mall for a California Town," *New York Times*, November 3, 2004.

Newman, William R. *Promethean Ambitions: Alchemy and the Quest to Perfect Nature.* Chicago: University of Chicago Press, 2004.

Orvell, Miles. *The Real Thing: Imitation and Authenticity in American Culture, 1880-1940.* Chapel Hill: University of North Carolina Press, 1989.

Paradis, James, and Thomas Postlewait, eds. *Victorian Science and Victorian Values: Literary Perspectives.* New Brunswick, NJ: Rutgers University Press, 1985.

Pohl, Otto. "A Defense from Portugal for the Noble Wine Cork," *New York Times*, October 14, 2001.

"Reproducing Amber," *Du Pont Magazine*, March 1921, pp. 10+.

Robertson, Sarah. "Faking It," *The Wall Street Journal*, February 28, 2003.

Rockwell, John. "Artifice Can Be Art's Ally as Well as Its Enemy," *New York Times*, March 12, 2004.

Rozhon, Tracie, and Rachel Thorner. "They Sell No Fake Before Its Time," *New York Times*, May 26, 2005.

Schröedinger, Erwin. *What Is Life?* Cambridge: Cambridge University Press, 1992.

Schwartz, Hillel. *The Culture of the Copy.* New York: Zone Books, 1998.

Tagliabue, John. "The End of Chocolate (as a Chocolatier Knows It)," *New York Times*, September 5, 2003.

Tompkins, Joshua. "When Technology Imitates Art," *New York Times*, July 22, 2004.

"Tortoise Shell and Shell Pyralin," *Du Pont Magazine*, April 1921, p. 13.

Trefethen, Jim. *Wooden Boat Renovation.* Camden, NJ: International Marine, 1993.

"Vermont Woodworkers a Certified Success," *AMC Outdoors*, October 2003, p. 22.

Vogel, Steven. "Unnatural Acts," *The Sciences*, July-August 1999.

Wood, Gaby. *Edison's Eve: A Magical History of the Quest for Mechanical Life.* New York: Alfred A. Knopf, 2002.

FURTHER ACKNOWLEDGMENTS

I wish to express my thanks to Rosalind Williams and Merritt Roe Smith for their excellent ideas and leads, to Jim Paradis for helping me find time to write while at MIT, to the superbly competent staff of the MIT library's many divisions and departments, and to my students in the MIT Graduate Program in Science Writing.

Warm thanks go as well to Philip Khoury, then dean of MIT's School of Humanities, Arts, and Social Sciences, and to Doron Weber, of the Alfred P. Sloan Foundation, for their generous support of *Faux Real.* To Jeff Robbins, my very capable editor, and the others at Joseph Henry Press, who helped bring this book into being; to Marcia Bartusiak for alerting me to Joseph Henry; to outside reviewers Frank E. Karasz and Edward J. Kramer for their many worthwhile suggestions; and to Susanne Martin and Shannon Larkin for their help and patience along the way.

Thanks to David for the book's title; to Rachele, Laird, Jessie, and Mom; and to Dad, who passed away just before I started on this book but who would have been interested in it, I am certain.

It is impossible for me to adequately thank Sarah for the light she has brought to my life.

Index

A

B

C

E

F

G

H

I

M

Q

R

T

U

V

W

Y

Z

Breinigsville, PA USA
06 August 2010
243084BV00001B/13/P